MyMathLab® NOTEBOOK

GEX Publishing Services

DEVELOPMENTAL MATHEMATICS: PREALGEBRA, INTRODUCTORY ALGEBRA AND INTERMEDIATE ALGEBRA

John Squires

Chattanooga State Community College

Karen Wyrick

Cleveland State Community College

PEARSON

Boston Columbus Indianapolis New York San Francisco Upper Saddle River
Amsterdam Cape Town Dubai London Madrid Milan Munich Paris Montreal Toronto
Delhi Mexico City São Paulo Sydney Hong Kong Seoul Singapore Taipei Tokyo

The author and publisher of this book have used their best efforts in preparing this book. These efforts include the development, research, and testing of the theories and programs to determine their effectiveness. The author and publisher make no warranty of any kind, expressed or implied, with regard to these programs or the documentation contained in this book. The author and publisher shall not be liable in any event for incidental or consequential damages in connection with, or arising out of, the furnishing, performance, or use of these programs.

Reproduced by Pearson from electronic files supplied by the author.

ISBN-13: 978-0-321-92311-0
ISBN-10: 0-321-92311-1

7

www.pearsonhighered.com

MyMathLab Developmental Mathematics: Prealgebra, Introductory Algebra, and Intermediate Algebra: A Modular Approach

John Squires, Karen Wyrick

Table of Contents

Name: _____   Date: _____

Instructor: _____   Section: _____

Whole Numbers
Topic 1.1 Whole Numbers

Vocabulary
whole number • standard form • place value • expanded notation

1. To better understand the value of each digit in a number, we can write it in

    _____.

Step-by-Step Video Notes
Watch the Step-by-Step Video lesson and complete the examples below.

Example	Notes
1. Consider the number 13,579. What digit is the tens place value? 7 What is the place value of the 3? _____ What is the actual value of the 3? 3000	
2. Write 4295 in expanded form. The 4 is in the thousands place value, so the actual value is 4000. What is the actual value of the 2? ☐ What is the actual value of the 9? ☐ What is the actual value of the 5? ☐ $4295 = 4000 + \square00 + \square0 + 5$ Answer:	

Example	Notes
4. Convert 1209 to words. Write the number in the first period on the left followed by the period name and a comma. one _____, Do this for the next period, but remember that the "ones" do not need the name of the period. Answer:	
6. Write twenty-four thousand, seven hundred twelve in standard form. Read the number from left to right. Write the number in the first period followed by a comma. ☐, Continue. Answer:	

Helpful Hints

When converting numbers to words or writing numbers in standard form, you always start at the left.

The word "and" is not included when reading or writing numbers.

Concept Check

1. How many place values are in a period?

Practice

Convert to words.

2. 5287

3. 321,608

Write in standard form.

4. two thousand six hundred four

5. twelve thousand, forty-eight

Whole Numbers
Topic 1.2 Rounding

Vocabulary

estimating   •   rounding   •   rounding down   •   rounding up

1. _____ is finding a number close to the exact number, but easier to work with.

Step-by-Step Video Notes
Watch the Step-by-Step Video lesson and complete the examples below.

Example	Notes
3. Round 506,243 to the nearest thousand. Underline the digit in the place value to which you are rounding. 50<u>6</u>,243 What is the digit to the right of the underlined digit? ☐ It is 4 or less, so the underlined digit stays the same. Replace the digit to the right of 6 with zeros. Answer: 506,☐☐☐	
4. Round 101,697 to the nearest hundred. 101,<u>6</u>97 Answer:	

Example	Notes
5. Round 1296 to the nearest ten. 12<u>9</u>6 Look at ☐, which is 5 or more so the under Lined digit increases by one. Note that when a 9 becomes a 10, the digit to the left of 9 also increases by one, so the 2 becomes a three. Answer: 13☐☐	
6. Round 449,985 to the nearest hundred. Answer:	

Helpful Hints

Remember to replace all the digits to the right of the digit you are rounding with zeros after rounding up or rounding down.

When a 9 is rounded up to ten, the digit to the left of the 9 must also be increased by one.

Concept Check
1. When rounding a number to the nearest place value, and the number to the right of the digit being rounded is 5, do you round up or round down?

Practice
Round the following numbers to the nearest place value indicated.

2. 4528 to the nearest ten

4. 894 to the nearest hundred

3. 23,179 to the nearest thousand

5. 49,926 to the nearest hundred

Whole Numbers
Topic 1.3 Adding Whole Numbers; Estimation

Vocabulary
addition   •   sum   •   estimation   •   commutative property of addition   •
addition property of zero   •   rounding   •   associative property of addition

1. _____ occurs when you combine numbers.

2. The property which states that changing the order when adding numbers does not change
    the sum is the _____.

Step-by-Step Video Notes
Watch the Step-by-Step Video lesson and complete the examples below.

Example	Notes
1. Find the sum of $5+6$. The sum is the answer to the addition, which is combining 5 and 6 and counting the total number of items. Answer:	
4. The statement $(3+5)+7 = 3+(5+7)$ is true. This shows that changing the grouping when adding numbers does not change the sum. Which property states this? Answer:	
7. Find the sum of $126+12$. Stack the numbers so that the digits in the same place value are lined up vertically. Begin at the right adding digits. $\begin{array}{r} 126 \\ +12 \\ \hline \end{array}$ Answer:	

Example	Notes
9. Estimate by rounding to the nearest hundred and comparing to the exact answer. $7026 + 13,479$ Round 7026 to the nearest hundred. 7☐00 Round 13,479 to the nearest hundred. 13,☐00 Find the sum of the rounded numbers. ☐ What is the exact answer? ☐ Answer:	

Helpful Hints

When adding numbers vertically, if the sum of any column is more than 9, you need to carry the tens place value number to the next column.

Estimation is a process that can help you check that your exact answer is close to the actual answer.

Concept Check

1. Can you name three words or phrases that indicate addition?

Practice

Find the following.

2. $135 + 22$

3. $111 + 246 + 3189$

4. 13 more than 8

5. sum of 24 and 57

Name: _____   Date: _____

Instructor: _____   Section: _____

Whole Numbers
Topic 1.4 Subtracting Whole Numbers

Vocabulary

difference   •   sum   •   estimation   •   subtraction

1. _____ of two numbers is taking away one number or quantity from another.

2. The _____ is the answer to a subtraction problem.

Step-by-Step Video Notes
Watch the Step-by-Step Video lesson and complete the examples below.

Example	Notes
3. Subtract $10-4$. Take 4 away from 10. $10-4=\square$ Check by adding. $4+\square=10$ Answer:	
5. Subtract $57-38$. Begin with the ones place and subtract the bottom digit from the top. Since the top digit is smaller than the bottom digit, borrow from the next place value. $\begin{array}{r} 57 \\ -38 \\ \hline \end{array}$ Now subtract. $\begin{array}{r} 4\ 17 \\ \cancel{5}\ \cancel{7} \\ -3\ 8 \\ \hline \square \end{array}$ Check by adding. Answer:	

Example	Notes
6. Subtract $232-141$. Check by adding.	
Answer:	
8. Estimate by rounding to the nearest thousand and comparing to the exact answer. $11,976-1245$ Round 11,976 to the nearest thousand. ☐ Round 1245 to the nearest thousand. ☐ Find the difference of the rounded numbers. Compare to the exact answer. Answer:	

Helpful Hints
Any number minus itself is zero.

Estimation is a process that can help you check that your exact answer is close.

Concept Check
1. Can you name three words or phrases that indicate subtraction?

Practice
Find the following.

2. $86-28$

3. $3045-2824$

4. 13 decreased by 8

5. 24 less than 52

Name: _____   Date: _____

Instructor: _____   Section: _____

Whole Numbers
Topic 1.5 Basic Problem Solving

Vocabulary

addition   •   perimeter   •   problem solving   •   translating

1. To find the _____ of a figure, add the lengths of all its sides.

2. The procedure for _____ involves creating a plan in symbols or words and performing calculations.

Step-by-Step Video Notes
Watch the Step-by-Step Video lesson and complete the examples below.

Example	Notes
1. Translate the sum of 5 and 3 to symbols. Enter the operation indicated by the word sum. 5 ☐ 3 Simplify. Answer:	
3. A rectangular garden measures 9 feet in length and is 5 feet wide. How many feet of fencing are needed to enclose the garden? Understand the problem. We are trying to find out how much _____ is needed for the garden. Create a plan. We need to find the _____ of the lengths of the four sides. Find the answer. Check. Answer:	

Example	Notes
4. A local community college requires 64 credit hours for an Associate Degree. Kylie earned 28 credits this past year. How many more credit hours does Kylie need in order to get her degree? Enter the operation indicated from the key word or phrase in the problem. credit hours required ☐ credit hours earned is credit hours needed Answer:	
5. On Monday, Alex opened a checking account With an initial deposit of $300. She bought groceries for $75, spent $25 on gas, and spent $20 on new clothes. How much money is in her account after these purchases? Answer:	

Helpful Hints

When solving a problem, there are key words or phrases which can be translated into an operation.

After obtaining an answer from your calculation, make sure to check that this answers the question which was asked in the problem.

Concept Check

1. What two steps need to be done before calculation in solving a problem?

Practice

Today is Dan's turn to bring water to soccer practice. Dan's mom put 24 water bottles in the cooler and his dad put 6 water bottles to the cooler. Dan's soccer team has 19 players, but 3 are not at practice today. Each player takes one water bottle at practice. Determine the following values.

2. The number of water bottles in cooler before practice.

4. The number of water bottles taken at practice.

3. The number of players at practice today.

5. The number of water bottles left in cooler after practice.

Whole Numbers
Topic 1.6 Multiplying Whole Numbers

Vocabulary
product • factors • addition • multiplication property of one • multiplication
commutative property of multiplication • associative property of multiplication
multiplication property of zero • distributive property of multiplication over addition

1. The _____ is the answer to a multiplication problem.

2. The _____ states that the changing the grouping when multiplying numbers
   does not change the product.

Step-by-Step Video Notes
Watch the Step-by-Step Video lesson and complete the examples below.

Example	Notes
3. For 3×4, identify the factors. 3, $\Box$ Find the product. $3\times4 = 3+3+3+3 = \Box$ Answer:	
4. Rewrite $8(3+6)$ using the Distributive Property. $8(3)+\Box(6)$ Simplify. Answer:	
7. Which property tells us that the following is true? $6\cdot25 = 25\cdot6$ Commutative Property, Associative Property, Distributive Property, Multiplication Property of Zero, Multiplication Property of One Answer:	

Example	Notes
12. Find the product $3 \cdot 248$.	

Multiply the bottom number by each digit on the top starting on the right.

$3 \times 8 = 24$

The 4 is written as the ones digit answer and the 2 is carried to the next place value.

$$\begin{array}{r} 2 \\ 248 \\ \underline{\times 3} \\ 4 \end{array}$$

Complete the multiplication.

Answer:

Helpful Hints
The Distributive Property can be used to make mental calculations easier.

When number is carried to the next place value in a multiplication problem, this number is added to the product of the next multiplication.

Concept Check
1. Can you name two words or phrases that indicate multiplication?

Practice
Rewrite using the Distributive Property. Simplify.

2. $3(9+7)$

Find the product.

4. $4 \cdot 164$

3. $7(5+2)$

5. $3 \cdot 192$

Whole Numbers
Topic 1.7 Dividing Whole Numbers

Vocabulary
division • quotient • dividend • divisor • long division • remainder
• divides exactly

1. The number you are dividing by is the _____.

2. The _____ is the answer to a division problem.

Step-by-Step Video Notes
Watch the Step-by-Step Video lesson and complete the examples below.

Example	Notes
2. Divide the following. $36 \div 4$ $$4 \cdot \square = 36, \text{ so } 36 \div 4 = \square$$ Answer:	
3. Divide the following. Use long division. $92 \div 4$ Answer:	

$$
\begin{array}{r}
2\,\square \\
4\,\overline{)\,9\;\;2} \\
-8 \\
\hline
\square\,2 \\
-\square\,2 \\
\hline
0
\end{array}
$$

Example	Notes
4. Divide the following. Use long division. 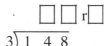 Answer:	

5. Find the quotient of 21 and 7.

$$7 \cdot \square = 21, \text{ so } 21 \div 7 = \square$$

Answer:

Helpful Hints

In order to divide, you need to have mastered your multiplication facts. You can check any division problem using multiplication.

Remember that any non-zero number divided by itself is 1, and any number divided by 1 is the number itself. Zero divided by any non-zero number is 0, and division by 0 is undefined.

When translating for division, be careful to write the numbers in the correct order.

Concept Check

1. When using the long division symbol, which number goes inside? Which goes outside?

Practice

Divide the following.

2. $48 \div 6$

3. $119 \div 4$

Find the quotient of the following.

4. 32 and 8

5. 84 and 12

Whole Numbers
Topic 1.8 More with Multiplying and Dividing

Vocabulary
division • quotient • estimation • remainder • product • multiplication

1. The _____ is the answer to a multiplication problem.

Step-by-Step Video Notes
Watch the Step-by-Step Video lesson and complete the examples below.

Example	Notes
2. Calculate the following product. $32 \cdot 125$ Set up the multiplication. It is preferable to put the larger number on top when setting up this problem. $\quad\quad 1\ 2\ 5$ $\times\quad\ \ 3\ 2$ $\quad\ \square\ 5\ 0$ $\ \square\square\square\ 0$ $\ \square\square\square\ 0$ Answer:	
3. Multiply. 260(400) $26 \cdot 4 = \square$ Attach the 3 ending zeros from the factors to the end of the product. $260(400) = \square,\square\square\square$ Answer:	

Example	Notes

4. Estimate by rounding to the nearest ten.

$7227(87) \approx 7230(\boxed{})$

$723 \cdot \boxed{} = \boxed{}$

Attach the 2 ending zeros from the factors to the end of the product

$7227(87) \approx \boxed{}\boxed{}\boxed{},\boxed{}\boxed{}\boxed{}$

5. Divide the following. Set up the division.

$1482 \div 12$

$$\boxed{}\boxed{}\boxed{}\,r\boxed{}$$
$$12\overline{)1 \quad 4 \quad 8 \quad 2}$$
$$-\boxed{}\boxed{}$$
$$\boxed{}\;8$$
$$-\boxed{}\boxed{}$$
$$\boxed{}\;2$$
$$-\boxed{}\boxed{}$$
$$\boxed{}$$

Helpful Hints
When multiplying or dividing whole numbers with several digits, be careful to neatly stack the numbers vertically so that digits in the same place value line up.

Concept Check
1. When multiplying numbers ending in zeros, do you have to write each entire number in columns before multiplying? What can you do to make the multiplication simpler?

Practice
Calculate the following products.
2. $31 \cdot 19$

3. $24 \cdot 225$

Divide the following.
4. $1768 \div 12$

5. $1876 \div 16$

Whole Numbers
Topic 1.9 Exponents

Vocabulary
exponent • base • squared • cubed • power of 1 • power of 0
exponential growth

1. A(n) _____ is a shortcut for repeated multiplication.

2. Any non-zero number raised to the _____ is equal to 1.

Step-by-Step Video Notes
Watch the Step-by-Step Video lesson and complete the examples below.

Example	Notes
1. Identify the base and exponent, then evaluate. 5^2 The base is $\square$. The exponent is $\square$. $5^2 = 5 \cdot 5 = \square$ Answer:	
4. Write the following using exponents, then evaluate. $(3)(3)(3)$ $(3)(3)(3) = 3^\square = \square$ Answer:	

Example	Notes
11. Evaluate. 31^0 Answer:	
12. Calculate 10^2, 10^3 and 10^6. $10^2 = \boxed{}$ $10^3 = \boxed{}$ $10^6 = \boxed{}$	

Helpful Hints

The base is the number, or factor, being multiplied by itself. The exponent is the number of times the base is used as a factor.

Any number raised to the power of 1 is the number itself. Any non-zero number raised to the power of 0 is equal to 1.

An exponent expression, for example, 7^2 does NOT mean $7 \cdot 2$. It means use 7 as a factor 2 times, in other words, $7 \cdot 7$. The value of the expression is 49.

Concept Check

1. Can you explain a simple rule for evaluating powers of ten? How does the exponent relate to the number of zeros in the standard form number? Give an example.

Practice

Identify the base and exponent, then evaluate.

2. 4^3

3. 9^2

Evaluate.

4. 10^7

5. 25^0

Whole Numbers
Topic 1.10 Order of Operations and Whole Numbers

Vocabulary
order of operations   •   PEMDAS   •   parentheses   •   exponents

1. When simplifying an expression using the order of operations, always evaluate what is inside _____ or other grouping symbols first.

Step-by-Step Video Notes
Watch the Step-by-Step Video lesson and complete the examples below.

Example	Notes
1. Calculate. $4 + 6 \cdot 3$ The operations in this expression are addition and multiplication. Multiply first, and then add. $4 + \square = \square$ Answer:	
3. Simplify by using the order of operations. $3^2 + 5 \cdot 4$ There is an exponent in the expression and multiplication. Evaluate the exponent then multiply. $\square + 5 \cdot 4 = \square + \square = \square$ Answer:	

Example	Notes
4. Simplify. $3(4-2)^2+5$ First simplify the operation in the _____, and then square that number. Next _____ by $\square$, then _____ $\square$. $3(\square)^2+5=3\cdot\square+5$ $=\square+5$ $=\square$ Answer:	

6. Simplify.

$$\frac{22+10}{4\cdot5-4}$$

Simplify the numerator. $\dfrac{\square}{4\cdot5-4}$

Simplify the denominator. $\dfrac{\square}{\square}$

Divide. $\square$

Helpful Hints
These memory tips might help you remember the order of operations. P E MD AS, and Please Excuse My Dear Aunt Sally.

Other symbols also act like parentheses: brackets such as [ ] and { }, and fraction bars. When you see a fraction bar, act like there are parentheses around the numerator and denominator.

Concept Check
1. The order of operations tells you to multiply and divide from left to right. Does the answer change if you multiply first if simplifying an expression such as $6\div3\cdot2$?

Practice

Evaluate.

2. $10-3\cdot2$

3. $4(9-6)^2-8$

Simplify.

4. $5^2+9\div3$

5. $\dfrac{14+13}{6\cdot2-3}$

Whole Numbers
Topic 1.11 More Problem Solving

Vocabulary

area   •   perimeter   •   problem solving   •   translating

1. The _____ of a rectangle is found by multiplying its length times its width.

Step-by-Step Video Notes
Watch the Step-by-Step Video lesson and complete the examples below.

Example	Notes
1. Translate twice the sum of 7 and 4 to symbols. Enter the appropriate operation symbol. $2\left(7\,\square\,4\right)$ Simplify. Answer:	
4. Find the area of a room that has a length of 21 feet and a width of 14 feet. Understand the problem. We need to find the _____ of the room. Create a plan. We will find the _____ by multiplying the length $\square$ feet by the width $\square$ feet. Find the answer. Check. Answer:	

Example	Notes
5. Desmond makes a salary of $39,000 per year. What is his monthly salary? We need to find _____. There are ☐ months in a year. We will _____ his yearly salary by ☐. Answer:	
6. The phone company charges $20 for installation and $25 a month for basic service. How much will Monique pay for one year of service with installation? Answer:	

Helpful Hints

When solving a problem, there are key words or phrases which can be translated into an operation.

After obtaining an answer from your calculation, make sure to check that this answers the question which was asked in the problem.

Concept Check

1. What two steps need to be done before calculation in solving a problem?

Practice

Find the area of the room with the following dimensions.

2. length of 21 feet and width of 12 feet

3. length of 10 feet and width of 8 feet

Joe earns $900 at his summer job where he works for three weeks, 20 hours each week.

4. How much did he earn each week?

5. How much did he earn per hour?

Integers
Topic 2.1 Understanding Integers

Vocabulary
integers   •   negatives   •   opposites   •   absolute value

1. _____ are whole numbers and their opposites.

2. The _____ of a number is the distance from zero on a number line.

3. _____ are two numbers that are the same distance from zero but appear on different sides of zero on the number line.

Step-by-Step Video Notes
Watch the Step-by-Step Video lesson and complete the examples below.

Example	Notes
1 & 2. Write the following in symbols. Find the opposite and simplify.	

	Symbols	Answer
The opposite of 8		
The opposite of −25		

3–5. Find the opposite.

16

−38

0

Example	Notes
6–8. Find the absolute value.	

$|4|$

4 is ☐ units from zero on the number line.

$|-13|$

$|0|$

Helpful Hints
You must use the " − " sign with a negative number to indicate that the number's value is less than 0. A positive number is written without a sign.

To find the opposite of a number, change the sign of the number. If the number is positive, its opposite is negative. If the number is negative, its opposite is positive. The opposite of 0 is 0.

To find the absolute value of a number, take the numerical part of the number, without the sign.

Concept Check
1. Why is it impossible for an absolute value to be negative?

Practice
Find the opposite.

2. −36

Evaluate.

4. $|-76|$

3. 94

5. $-|48|$

Name: _____   Date: _____

Instructor: _____   Section: _____

Integers
Topic 2.2 Adding Integers

Vocabulary
opposites   •   additive inverses   •   absolute values

1. _____ are also known as additive inverses.

Step-by-Step Video Notes
Watch the Step-by-Step Video lesson and complete the examples below.

Example	**Notes**
1 & 2. Add. $$32+41$$ To add numbers with the same sign, add the absolute values (or numerical parts). The sign of the sum is the same as the sign of the original numbers. $$\begin{array}{r} 32 \\ +\ 41 \\ \hline \square \end{array}$$ $$-6+(-17)$$	
4. Add $-18+7$. To add two numbers with different signs, subtract the absolute values (or numerical parts). The sign of the answer is the same as the sign of the number with the larger absolute value. Answer:	

Example	Notes
5. Add $6+(-25)$. Answer:	
6. Add $3+(-5)+(-2)$. Add from left to right. Answer:	

Helpful Hints

When adding two numbers with the same sign, the sign of the sum is the same as the sign of the original numbers. When adding two positive numbers, the answer will be positive. When adding two negative numbers, the answer will be negative.

The sum of a number and its opposite is zero.

When adding two numbers with different signs, the sign of the answer is the same as the sign of the number with the larger absolute value (the number that is farther from 0 on a number line).

Concept Check

1. Ziva wants to find the sum of $4+(-18)+12+(-13)+5$ by first adding $4+12+5$, then adding $(-18)+(-13)$, and then adding to two results. Will her answer be correct? Explain.

Practice
Add.

2. $-8+(-7)$

3. $23+(-12)$

4. $21+19$

5. $5+(-13)+17+(-6)$

Integers
Topic 2.3 Subtracting Integers

Vocabulary
adding the opposite   •   additive inverses   •   absolute value   •   subtracting

1. Subtraction is "_____." If $5-3=2$, then $5+(-3)=2$.

Step-by-Step Video Notes
Watch the Step-by-Step Video lesson and complete the examples below.

Example	**Notes**
1. Subtract $-8-(-3)$. Leave the first number alone. Change the minus sign to a plus sign. Change the sign of the number being subtracted. Add the two numbers using the rules of addition. Answer:	
2. Subtract $-5-(-9)$. Use the Leave, Change, Change method. $-5+\left(\boxed{}\right)=\boxed{}$ Answer:	

Example	Notes
3. Subtract $-61 - 23$.	
Answer:	
4. Today at 4:00 p.m., the temperature outside was $-4°$ F. Three hours later, the temperature had dropped 9 more degrees. What was the temperature at 7:00 p.m.?	
Answer:	

Helpful Hints

To subtract, *leave* the first number alone, *change* the subtraction symbol to an addition symbol, and *change* the sign of the number being subtracted to its opposite. Then follow the rules for addition.

Subtracting a negative number is the same as adding a positive number. Subtracting a positive number is the same as adding a negative number.

Concept Check
1. How does the concept of absolute value help to explain the statement "Subtracting a negative number is the same as adding a positive number?" Use $2 - (-6)$ as an example.

Practice

Subtract.
2. $-10 - 7$

3. $31 - (-19)$

Subtract.
4. $-7 - (-9)$

5. The temperature this morning was $-2°$. By 8:00 p.m. the temperature dropped 8 more degrees. What was the temperature at 8:00 p.m.?

Name: _____   Date: _____

Instructor: _____   Section: _____

Integers
Topic 2.4 Multiplying and Dividing Integers

Vocabulary
absolute value   •   negative   •   positive   •   even   •   odd

1. When multiplying two numbers with the same sign, the answer is _____.

2. When multiplying two numbers with different signs, the answer is _____.

Step-by-Step Video Notes
Watch the Step-by-Step Video lesson and complete the examples below.

Example	**Notes**
1. Multiply. $(-12)(-9)$ Multiply the numerical parts. Determine the sign of the answer. Answer:	
2. Multiply. $(7)(-4)$ Answer:	
5. Multiply $(-2)(-3)(-4)(5)$. Answer:	

Example	Notes
8. Divide $3\overline{)-48}$. Divide the numerical parts. $3\overline{)48}$ Answer:	
10. Divide $\dfrac{-28}{-7}$. Answer:	

Helpful Hints
When multiplying an even number of negative factors, the product is positive. When multiplying an odd number of negative factors, the product is negative.

The sign rules for division are exactly the same as for multiplication.

Concept Check
1. Why will the product of any two negative integers be larger than the product of any three negative integers?

Practice

Multiply.

2. $(-6)(-5)$

3. $(-3)(7)(-2)(-3)(-1)$

Divide.

4. $-19\overline{)95}$

5. $\dfrac{-72}{18}$

Integers
Topic 2.5 Exponents and Integers

Vocabulary

multiplication • exponent • base • order of operations • dividing

1. In an exponential expression, the _____ is the number of times the number is being multiplied.

2. In an exponential expression, the _____ is the number being multiplied.

Step-by-Step Video Notes
Watch the Step-by-Step Video lesson and complete the examples below.

Example	**Notes**
1 & 2. Write in exponential form. $7 \cdot 7 \cdot 7 \cdot 7 \cdot 7$ The base is $\square$. The base is being multiplied $\square$ times. $(-8)(-8)(-8)(-8)(-8)(-8)$	
5–8. Evaluate. $(-4)^2$ $\qquad$ $(-4)^3$ -4^2 $\qquad$ -4^3	

Example	Notes
10. Evaluate $(-23)^1$.	
Answer:	
12. Evaluate -10^9.	
Answer:	

Helpful Hints

When the base is negative, be especially careful in determining the sign of the answer. Be careful to recognize when a negative sign is part of the base. For example, in -7^3, the negative sign is not included; but in $(-7)^3$, it is.

A negative base raised to an even power is positive. A negative base raised to an odd power is negative.

Any number raised to the power of 1 is itself.

When 10 is raised to a power, the exponent gives the number of trailing zeros in the answer.

Concept Check

1. Which is a greater number, $(-9)^{15}$ or $(-9)^4$? Explain.

Practice

2. Write in exponential form.

$(-13)(-13)(-13)(-13)$

3. Evaluate $(-7)^2$.

4. Evaluate $-(-3)^3$.

5. Evaluate $(-10)^6$.

Name: _____   Date: _____

Instructor: _____   Section: _____

Integers
Topic 2.6 Order of Operations and Integers

Vocabulary
multiplying　　•　　dividing　　•　　order of operations　　•　　grouping symbols

1.　When evaluating an expression with more than one operation, use _____.

Step-by-Step Video Notes
Watch the Step-by-Step Video lesson and complete the examples below.

Example	**Notes**
1.　Simplify $(-5)^2 + 3^2$. P: Parentheses E: Exponents $\quad (-5)^2 \;+\; 3^2$ $(\square)(\square)\;(\square)(\square)$ $\quad \square \;+\; \square$ <u>MD</u>: Multiply and Divide <u>AS</u>: Add and Subtract Answer:	
3.　Simplify $(-6+4)^3 \div (-2)$. P: Parentheses E: Exponents <u>MD</u>: Multiply and Divide <u>AS</u>: Add and Subtract Answer:	

Example	Notes
4. Simplify $\lvert 3-5 \rvert + 4 \div 2 \cdot 2$. Absolute value bars can also act as grouping symbols. Answer:	
5. Simplify $\dfrac{4^2 + (2)(-3)}{-4+6-7}$. Answer:	

Helpful Hints

Note that { }, [ ], $\lvert\ \rvert$, and fraction bars can also act as grouping symbols.

Use the phrase "**P**lease **E**xcuse **M**y **D**ear **A**unt **S**ally" to remember the order of operations, P E MD AS.

Concept Check

1. To simplify $-9 \cdot 6 \div \lvert 2-5 \rvert$, which operation will you perform first?

Practice

Simplify.

2. $(-8)^2 + (-3)^2$

3. $4^3 \div (-4+6)^3$

4. $\lvert 1-8 \rvert^2 - (24 \div 2) + 6$

5. $\dfrac{7^2 - 9 \cdot 5}{(-6)^2 + 5(-7)}$

Introduction to Algebra
Topic 3.1 Variables and Expressions

Vocabulary
variable   •   term   •   expression   •   algebraic expression   •   coefficient

1. A(n) _____ is a number, variable, or product of numbers and/or variables.

2. A(n) _____ is a symbol (usually a letter) used to represent an unknown number.

Step-by-Step Video Notes
Watch the Step-by-Step Video lesson and complete the examples below.

Example	Notes
2. Identify the coefficient and the variable(s) in $-8abc$. ☐ is the coefficient. Identify the variable(s). Answer:	
3 & 4. Decide if each of the following is a term, an algebraic expression, or both. $57g$ $4t - 8v$	
5. Evaluate the algebraic expression $3n$ for $n = 7$. Replace the variable with the given value. $3\left(\boxed{}\right)$ Multiply. Answer:	

Example	Notes
6. Evaluate the algebraic expression $x + y - 5$ for $x = 2$ and $y = 9$. Replace the variables with the given values, and simplify using the order of operations. 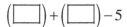 Answer:	
9. Translate "twice the sum of three and a number" from words to symbols. Use n as the variable. Answer:	

Helpful Hints

An algebraic expression is a collection of one or more terms separated by a " + " or a " −." They can contain one or more variables.

If a term is negative, the negative sign is part of the coefficient.

When writing variables, do not use the letter O as a variable, because it can be confused with the number zero (0).

Concept Check

1. How many terms are in the expression $7 + 2p$? Is it an algebraic expression?

Practice

Evaluate the expression for the given values of the variables.

2. $-6(x - y)$ for $x = -3$ and $y = 4$

3. $8m + 4q - 9$ for $m = 5$ and $q = -7$

4. Identify the terms, coefficients, and variables in $9 - 3x + w$.

5. Translate "double the difference a number and eight" from words to symbols. Use n as the variable.

Introduction to Algebra
Topic 3.2 Like Terms

Vocabulary
coefficients   •   like terms   •   combining like terms   •   terms

1. Terms that have the same variables raised to the same powers are _____.

2. _____ are the number factor in a term.

Step-by-Step Video Notes
Watch the Step-by-Step Video lesson and complete the examples below.

Example	**Notes**

2 – 4. Identify the like terms in each list.
 	
  $4a,\ 9,\ 13a$
 	
  $-5st,\ \ st,\ \ 7t,\ \ 18st,\ \ 9s$
 	
  $3x^2,\ 2,\ -5x^2, 9x^2, -7$	

6. Combine like terms.
 	
  $7m + 3m - 6m$
 	
  $7m + 3m - 6m = \boxed{}m - \boxed{}m = \boxed{}m$
 	
  Answer:	

8. Simplify $6c + 3c - 13c + c$ by combining like terms. Answer:	

Example	Notes
10. Simplify $2x + (-3y) - 4 + 6y + 1$ by combining like terms. Answer:	

Helpful Hints

To combine like terms, add or subtract the numerical coefficients, or number parts. The variable part remains the same.

Anytime a single variable is shown the coefficient is understood to be 1. For example, $x = 1x$.

Terms do not need to have variables to be like terms. Constant terms are like terms, too.

Concept Check

1. Are y^2 and $2y$ like terms? Explain.

Practice

Simplify by combining like terms.

2. $-6x + 3x + 5x$

4. $4w - 5z + 7w + 6$

3. $8a + 3a - a$

5. $7q + 4s + (-12q) + s - 3$

Introduction to Algebra
Topic 3.3 Distributing

Vocabulary
like terms   •   order of operations   •   multiplication   •   Distributive Property

1. The _____ states that if a, b, and c are numbers or variables, then
   $a(b+c) = ab + ac$ and $a(b-c) = ab - ac$.

Step-by-Step Video Notes
Watch the Step-by-Step Video lesson and complete the examples below.

Example	Notes
1 & 2. Multiply using the Distributive Property. $4(x+3)$ Multiply the 4 by each term in the parentheses. $4(x+3) = \square(x) + \square(3)$ $-2(a-9)$	
4. Multiply $(4-a)3$ using the Distributive Property. $(4-a)3 = 4(\square) - a(\square)$ Simplify each term. Answer:	

Example	Notes
5 & 6. Multiply using the Distributive Property. $$7(4x-3y)$$ $$2(3m+5n-1)+6$$	

Helpful Hints.

The order of operations indicates that we perform the operation inside the parentheses first, but this is not always possible. When this occurs, we use the Distributive Property to rewrite the expression without parentheses.

The Distributive Property can be used when there are more than two terms in the parentheses.

Only use the Distributive Property if the numbers or variables inside the parentheses are separated by a "+" or "−" sign.

Concept Check

1. Does the expression $4(xy)$ require the Distributive Property to simplify?

Practice

Multiply using the Distributive Property.

2. $7(x-3)$

3. $(8+s)6$

4. $-5(2x+9y)$

5. $-3+4(6m-2n+3)$

Introduction to Algebra
Topic 3.4 Simplifying Expressions

Vocabulary
term   •   variable   •   expression   •   algebraic expression

1. A(n) _____ includes one or more terms separated with a " + " or a " − ".

2. A(n) _____ is an expression that contains one or more variables.

Step-by-Step Video Notes
Watch the Step-by-Step Video lesson and complete the examples below.

Example	**Notes**
1. Simplify $5a - 3 + 4(a + 7)$. Use the Distributive Property to remove the parentheses. $5a - 3 + 4(\square) + 4(\square)$ Combine like terms. $5a - 3 + 4a + 28 = \square a + \square$ Answer:	
2 & 3 Simplify. $-(4x - 3y)$ $-(5 + 6a)$	

Example	Notes
5. Simplify. $2\left[3(6x-5)\right]$ Answer:	
6. Simplify. $3x+7x(5-3)$ Answer:	

Helpful Hints.
To simplify an expression with a negative sign in front of parentheses, remove the parentheses by multiplying what is inside the parentheses by -1.

Concept Check
1. Why are the parentheses removed before combining like terms when simplifying algebraic expressions?

Practice
Simplify.
2. $7+3(2x-3)$

4. $-(4x+3)$

3. $-3\left[2(2x-7)\right]$

5. $2(y+3)-7(5+3y)$

Introduction to Algebra
Topic 3.5 Translating Words into Symbols

Vocabulary

difference   •   quotient   •   Commutative Property

1. The order in which you write a subtraction or division expression matters, because the _____ does not work for subtraction or division.

2. Phrases such as "a number decreased by 6," "6 fewer than a number," and "the _____ between a number an 6" all indicate subtraction.

Step-by-Step Video Notes
Watch the Step-by-Step Video lesson and complete the examples below.

Example	Notes
1–3. Translate into an algebraic expression. The sum of 8 and a number Triple a number The product of 13 and a number	
4–7. Translate into an algebraic expression. Five less than twelve Twelve less than five Fifty divided by one One divided by fifty	

Example	Notes
11–13. Write as an algebraic expression. Use parentheses if necessary. Seven more than double a number x A number x is tripled and then increased by 8 The quotient of three more than a number x and two	
14. Use an expression to describe the measure of each angle. The measure of the second angle of a triangle is double the measure of the first angle, and the third angle is 15° more than the measure of the second angle.	

Helpful Hints

The order of the terms is important in subtraction and division expressions.

Use parentheses when writing expressions to be sure certain operations are performed first.

Concept Check

1. What are some words or phrases in an expression that indicate addition?

Practice

Translate into an algebraic expression.

2. A number n increased by seven

3. The quotient of 24 and some number y

4. Triple the difference of a and b

5. 100 less than the product of 3 and x

Equations
Topic 4.1 Equations and Solutions

Vocabulary
equation • linear equation • solution of a linear equation • equivalent linear equations

1. A(n) _____ is a mathematical statement that two expressions are equal.

2. A(n) _____ is the number that, when substituted for the variable, makes the equation true.

Step-by-Step Video Notes
Watch the Step-by-Step Video lesson and complete the examples below.

Example	Notes
1–4. Identify each as an expression or equation. $3(x-6)$ $4x+3=7$ $2x+4y-8$ $7(x-4)+3=4x-2$	
5. Determine if 1 is a solution of $2x-3=1$. Substitute 1 into the equation for x. $2(\square)-3=1$ Simplify the left side of the equation according to the order of operations. Determine if the result is a true statement. Answer:	

Example	Notes
7. Determine if 6 is a solution to $x - 2 = 4$, graphically. 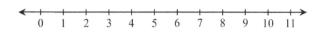 Answer:	
8. Determine if $6 - 4 = x$ and $2x + 5 = 9$ are equivalent equations. Solve the equation $6 - 4 = x$. Substitute the result into the other question and determine if the result is a true statement. Answer:	

Helpful Hints

An expression can be a sum or difference of terms, does not have an equal sign, and may be simplified. An equation is a statement that two expressions are equal, has an equal sign, and may be solved.

A linear equation may have one solution, no solution, or all real numbers may be the solution.

Concept Check

1. Use $x = 4$ to show that the equations $3x - 7 = 8$ and $5x + 2 = 22$ are not equivalent.

Practice

2. Identify each as an expression or equation.

   $9x + 16 = -20$

   $x - 2y + 4z - 10$

3. Determine if -4 is a solution to $3x = 6(x + 2)$.

4. Determine if 5 is a solution to $3x - 4 = 10$.

5. Determine if $x + 5 = 2(x + 3)$, $x + 3 = 4$, and $x = 1$ are equivalent equations.

Equations
Topic 4.2 Solving Equations by Adding or Subtracting

Vocabulary
addition • reverse operation • solution of an equation • Addition Property of Equality

1. The _____ states that adding the same number to both sides of an equation does not change the solution of the equation.

Step-by-Step Video Notes
Watch the Step-by-Step Video lesson and complete the examples below.

Example	**Notes**
1. Use the Addition Property of Equality to solve the equation $x - 3 = 8$. $x - 3 = 8$ $x - 3 + \Box = 8 + \Box$ $x = \Box$ Check your solution. Answer:	
2 & 3. Solve each equation by adding or subtracting. Check your solution. $x - 5 = 3$ $x + 2 = -9$ $x + 2 = -9$ $-\Box \quad -\Box$	

Example	Notes
5. Solve $2 = x - 7$ by adding or subtracting. Check your solution. Answer:	
6. Solve for x in the figure, if the perimeter of the figure is 15 inches. 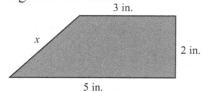 Answer:	

Helpful Hints

Remember to always keep the equation balanced. If you add a value to one side of the equation, you must also add the same value to the other side. If you subtract a value from one side of the equation, you must also subtract the same value from the other side.

The Addition Property of Equality in symbols states that if $a = b$, and c is any number, then $a + c = b + c$.

Concept Check
1. Explain how the Addition Property of Equality is used to solve $x + 2 = 1$. How does this differ from how it would be used to solve $x - 2 = 1$?

Practice

2. Solve $x + 6 = -3$ by adding or subtracting. Check your solution.

4. Solve $5 = x - 7$ by adding or subtracting. Check your solution.

3. Solve $x - 4 = 14$ by adding or subtracting. Check your solution.

5. Solve for x in the figure, if the perimeter of the figure is 24 inches.

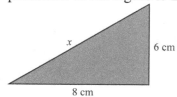

Equations
Topic 4.3 Solving Equations by Multiplying or Dividing

Vocabulary
division   •   multiplication   •   reverse operation   •   Multiplication Property of Equality

1. The _____ states that multiplying both sides of an equation by the same
    number (except zero) does not change the solution of the equation.

Step-by-Step Video Notes
Watch the Step-by-Step Video lesson and complete the examples below.

Example	Notes
1. Solve $3x = 21$ for x. Check your solution. Identify the variable. The goal is to get this variable alone on one side of the equation and the numbers on the other side. Used the Multiplication Property to divide each side of the equation by $\Box$ and simplify. $\dfrac{3x}{\Box} = \dfrac{21}{\Box}$ $x = \Box$ Check your solution. Answer:	
4. Solve $\dfrac{x}{-4} = 3$ by multiplying or dividing. Check your solution. $\dfrac{(\Box)\,x}{-4} = (\Box)\,3$ Answer:	

Example	Notes
5. Find the length of the rectangle if the area of the rectangle is 24 ft^2. The area A of a rectangle is given by the formula $A = lw$, where l is the length and w is the width. 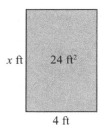 Answer:	

Helpful Hints

The Multiplication Property of Equality in symbols states that, for real numbers a, b, and c, when $c \neq 0$, if $a = b$, then $ca = cb$. Since division can be performed by multiplying by the reciprocal, this property works for division as well; if $a = b$, then $\dfrac{a}{c} = \dfrac{b}{c}$.

Concept Check

1. Explain how the Multiplication Property of Equality is used to solve $3x = 12$. How does this differ from how it would be used to solve $\dfrac{x}{3} = 12$?

Practice

2. Solve $\dfrac{x}{-4} = 16$ by multiplying or dividing. Check your solution.

4. Solve $\dfrac{x}{10} = 9$ by multiplying or dividing. Check your solution.

3. Solve $-5x = 35$ by multiplying or dividing. Check your solution.

5. Find the width of the rectangle if the area of the rectangle is 35 in.^2.

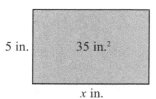

Equations
Topic 4.4 Solving Equations - Two Steps

Vocabulary
equation • Addition Property of Equality • solve • Multiplication Property of Equality

1. The _____ states that you can multiply (or divide) both sides of an equation by the same non-zero value without changing the solution.

Step-by-Step Video Notes
Watch the Step-by-Step Video lesson and complete the examples below.

Example	Notes
1. Solve for x. Check your solution. $$3x + 4 = 10$$ Use the Addition Property of Equality to get the variable term alone on one side of the equation. $$3x + 4 = 10$$ $$-\Box \quad -\Box$$ Simplify. Use the Multiplication Property of Equality to get the variable alone on one side of the equation. $$\frac{3x}{\Box} = \frac{6}{\Box}$$ Simplify. $$x = \Box$$ Check the solution. Answer:	

Example	Notes
3. Solve for x. Check your solution. $9 = -7 + 8x$ Use the Addition Property to add $\square$ to both sides. Use the Multiplication Property to divide both sides by $\square$. Answer:	
4. Solve for x. Check your solution. $\dfrac{x}{3} - 6 = 4$ Answer:	

Helpful Hints

To solve an equation of the form $ax + b = c$, where a, b, and c are real numbers, first get the variable term alone on one side of the equation. Then, get the variable alone on one side of the equation and simplify if needed. Finally, check the solution.

Concept Check

1. Describe the steps used to solve $3x + 5 = 16$.

Practice

Solve for x. Check your solution.

2. $4x + 7 = 19$ 4. $2x - 3 = 5$

3. $-3 = \dfrac{x}{7} - 8$ 5. $\dfrac{x}{3} + 4 = -2$

Equations
Topic 4.5 Solving Equations - Multiple Steps

Vocabulary
equation • addition property of equality • division • multiplication property of equality

1. The _____ states that if both sides of an equation are multiplied by the same non-zero value, the solution does not change.

Step-by-Step Video Notes
Watch the Step-by-Step Video lesson and complete the examples below.

Example	**Notes**
1. Solve $5x + 4 - 3x = -8$. Simplify each side by combining like terms. $5x + 4 - 3x = -8$ $\square + 4 = -8$ Get the variable terms on one side and the number terms on the other side. $2x + 4 = -8$ $-\square -\square$ Get the variable alone on one side. $\dfrac{2x}{\square} = \dfrac{\square}{\square}$ Simplify. Check. Answer:	

Example	Notes
2. Solve $2x - 5 = 6x + 7$. Answer:	
3. Solve $3(x + 5) = -6$. Remove parentheses. Answer:	
4. Solve $5(2x - 1) + 16 = -3(x + 5)$. Answer:	

Helpful Hints

Removing the variable term with the smaller coefficient from both sides of the equation will result in a positive variable term on one side of the equation.

Concept Check

1. List the six steps that may be needed in solving a multiple step equation.

Practice

Solve.

2. $2(x + 1) = -4$

4. $5x + 8 = 3x + 12$

3. $7x + 1 - 2x = -9$

5. $3(3x - 1) - 4 = 4(x + 2)$

Equations
Topic 4.6 Translating Words into Equations

Vocabulary
variable   •   equation   •   expression

1. A(n) _____ is a letter or symbol that is used to represent an unknown quantity.

2. A(n) _____ is a mathematical statement that two expressions are equal.

Step-by-Step Video Notes
Watch the Step-by-Step Video lesson and complete the examples below.

Example	Notes
1. Translate the following into an equation. Let n represent the number. Seven times a number is fourteen. Remember to go one word or phrase at a time when translating words into equations. Seven times a number is fourteen. ☐ • ☐ = ☐ Answer:	
2. Translate the following into an equation. Do not solve. Five more than six times a number is three hundred five. Answer:	

Example	Notes
3. Translate the following into an equation. Do not solve. The larger of two numbers is three more than twice the smaller number. The sum of the numbers is thirty-nine. Write an expression to represent each unknown quantity in terms of the chosen variable. Let n stand for the smaller number. An expression for the larger number is _____. Answer:	
4. Translate the following into an equation. Do not solve. The annual snowfall in Juneau, Alaska is 105 inches. This is 20 inches less than three times the annual snowfall in Boston, Massachusetts. Answer:	

Helpful Hints
When writing an equation from a word problem, choose a variable to represent an unknown quantity, write an expression to represent each unknown quantity in terms of the variable, and use a given relationship in the problem or an appropriate formula to write an equation.

Concept Check
1. What words or phrases are typically associated with multiplication? Division? Addition? Subtraction?

Practice
Translate into an equation. Do not solve.

2. A number n increased by three is nine.

3. A number divided by 4 is 16.

4. Twice the difference of a and 7 is eight.

5. 100 less than the product of 3 and x is 5.

Equations
Topic 4.7 Applications of Equations

Vocabulary

variable   •   equation   •   algebraic expression

1. A(n) _____ is a collection of one or more terms separated by a "+" or a "−."

2. A(n) _____ is a mathematical statement that two expressions are equal.

Step-by-Step Video Notes
Watch the Step-by-Step Video lesson and complete the examples below.

Example	Notes
1. Translate the following into an equation and solve. Five more than six times a number is one hundred eighty-five. Understand the problem. Choose a variable to represent an unknown quantity. 	

Five	more than	six times a number	is	one hundred eighty-five.
☐	+	☐	=	☐

Solve the equation.

Answer:

Example	Notes
2. Translate the following into an equation and solve. A proposed traffic law gives a fine for speeding of fifteen dollars plus ten dollars for every mile per hour of speed over the speed limit. If the ticket costs a driver $145, find how much over the speed limit the driver was driving. Answer:	
3. Translate the following into an equation and solve. A large bluefin tuna caught in the Atlantic Ocean weighed 1404 pounds. This is 246 pounds less than five times the weight of a large bluefin tuna caught in the Pacific Ocean. Find the weight of the tuna caught in the Pacific. Answer:	

Helpful Hints

Remember to write the answer in the context of the problem.

Concept Check

1. Sarah translated "Five less than a number is twelve" incorrectly as $5 - n = 12$. Explain her error and correct it.

Practice

Translate the following into an equation and solve.

2. Seven more than three times a number is twenty-five.

3. Sixteen is fourteen less than ten times a number.

4. The annual snowfall in one city is eight inches more than three times the snowfall in another city. If the snowfall in the first city is 47 inches, what is the annual snowfall in the second city?

Factors and Fractions
Topic 5.1 Factors

Vocabulary
factor   •   divisibility rule(s)   •   common factor   •   greatest common factor (GCF)

1. When two or more numbers are multiplied, each is called a _____.

2. The _____ of two or more terms is the largest number and/or variable that divides exactly into each of the terms.

Step-by-Step Video Notes
Watch the Step-by-Step Video lesson and complete the examples below.

Example	Notes
3. Find the greatest common factor, or GCF, of 15 and 24. List the factors of 15. 1, ☐, ☐, 15 List the factors of 24. 1, 2, ☐, ☐, ☐, 8, ☐, ☐ Identify the factors common to both lists. 1, ☐ Choose the largest of these. ☐ Answer:	
6. Find the GCF of y, y^4, and y^7. Choose the variable which is common to all the terms in the list and is raised to the smallest power. Answer:	

Example	Notes
7. Find the GCF of $9x^2$ and $15x^3$.	
Find the GCF of the coefficients or numerical parts.	
Find the GCF of the variable factors.	
Answer:	

8. List four ways to factor 18.

Answer:

Helpful Hints

A factor is a whole number or variable that divides exactly into another term.

There can be one or more common factors between two or more numbers, but there is only one greatest common factor (GCF).

The word "factor" can be either a noun or verb. Factors are the numbers being multiplied, and to factor a number means to write it as a product.

The greatest common factor (GCF) of two terms may be 1.

Concept Check

1. How many common factors are there between 24 and 96? Which is the GCF?

Practice

2. Find the GCF of 30 and 85.

4. Find the GCF of $35n^4$ and $49n^3$.

3. Find the GCF of x^5, x^3, and x^8.

5. List four ways to factor 99.

Factors and Fractions
Topic 5.2 Prime Factorization

Vocabulary
factor   •   prime number   •   composite number   •   prime factorization

1. A _____ is a whole number greater than 1 that has exactly two factors, the
   number itself and 1.

Step-by-Step Video Notes
Watch the Step-by-Step Video lesson and complete the examples below.

Example	Notes
1. Find the factors of 36. 1, 2, 3, 4, 6, ☐, ☐, ☐, 36	
3. Find the prime factorization of 24.	

Example	Notes
4. Find the prime factorization of 60.	

Helpful Hints

There is more than one way to write the factorization of a composite number. However, if all factors are rewritten as products of prime numbers, there is only one answer.

In the first step of rewriting a composite number as a product of two factors, there is often more than one combination of two factors that can be selected. However, continuing the steps of prime factorization will lead to the same final answer.

Concept Check

1. In finding the prime factorization of 72, if 72 is written as $6 \cdot 12$, what is the following step? And the next step?

Practice

2. Find the prime factorization of 56.

3. Find the prime factorization of 98.

4. Find the prime factorization of 210.

Name: _____   Date: _____

Instructor: _____   Section: _____

Factors and Fractions
Topic 5.3 Understanding Fractions

Vocabulary
fraction   •   denominator   •   undefined   •   improper fraction
proper fraction   •   numerator

1. A(n) _____ is the bottom number in a fraction and indicates the number of parts in the whole.

2. A(n) _____ is a fraction where the numerator is greater than or equal to the denominator.

Step-by-Step Video Notes
Watch the Step-by-Step Video lesson and complete the examples below.

Example	Notes
2a. How many parts are shaded in the diagram? How many total parts are in the whole diagram? □ Write a fraction that represents the shaded part. Answer:	
2b. Write a fraction that represents the shaded part.	

Example	Notes
3. The numerator of $\frac{4}{9}$ is ___ the denominator. Identify this fraction as proper or improper. Answer:	
6. Identify $\frac{3}{3}$ as proper or improper. Answer:	

Helpful Hints

When the numerator is equal to the denominator of a fraction, this is an improper fraction that is equal to 1.

A fraction with a numerator of zero is equal to zero; a fraction with a denominator of zero is undefined.

Concept Check

1. There are 9 boys in the class of 20 students. Represent this as a fraction and state why this is a proper fraction.

Practice

2. Write a fraction that represents the shaded part.

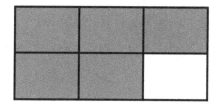

3. Identify $\frac{7}{7}$ as proper or improper.

4. Identify $\frac{2}{9}$ as proper or improper.

5. Identify $\frac{0}{5}$ as proper or improper.

Name: _____   Date: _____

Instructor: _____   Section: _____

Factors and Fractions
Topic 5.4 Simplifying Fractions - GCF and Factors Method

Vocabulary
equivalent fractions   •   simplest form   •   greatest common factor

1. A fraction is said to be in _____ if there is no common factor other than 1 that divides exactly into the numerator and the denominator.

2. _____ are fractions that represent the same value.

Step-by-Step Video Notes
Watch the Step-by-Step Video lesson and complete the examples below.

Example	Notes
1. Write $\dfrac{9}{15}$ in simplest form using the GCF method. Find the GCF of 9 and 15. GCF = ☐ Divide the numerator and the denominator by the GCF. Answer:	
2. Write $\dfrac{54}{90}$ in simplest form using the GCF method. Answer:	

Example	Notes
3. Write $\dfrac{20x}{32x}$ in simplest form using the factors method. $\dfrac{20x}{32x} = \dfrac{20x \div \Box}{32x \div \Box} = \dfrac{10x \div \Box}{16x \div \Box}$ Answer:	
5. Write $\dfrac{-30}{105}$ in simplest form using the factors method. Answer:	

Helpful Hints

Dividing the numerator and denominator by the same non-zero number does not change the value of a fraction. However, doing this gives us an equivalent fraction.

When looking for common factors, start by using small prime numbers like 2, 3, or 5.

Concept Check

1. Which method would you use for simplifying $\dfrac{8}{56}$? For simplifying $\dfrac{48}{175}$? Explain.

Practice

Write in simplest form using the GCF method.

2. $\dfrac{24}{27}$

3. $\dfrac{35s}{175s}$

Write in simplest form using the factors method.

4. $\dfrac{-160}{384}$

5. $\dfrac{120y}{210y}$

Factors and Fractions
Topic 5.5 Simplifying Fractions - Prime Factors Method

Vocabulary
prime factorization   •   prime factors method   •   equivalent fraction

1. To simplify a fraction using the _____, write the numerator and denominator as a product of prime numbers and variables without using exponents. Then, divide the numerator and denominator by common factors. Finally, multiply the remaining factors to get the simplest form.

Step-by-Step Video Notes
Watch the Step-by-Step Video lesson and complete the examples below.

Example	Notes
1. Write $\dfrac{24}{36}$ in simplest form using the prime factors method.	

Write the numerator as a product of prime numbers.

$$24 = 2 \cdot 2 \cdot \square \cdot \square$$

Write the denominator as a product of prime numbers.

$$36 = \square \cdot \square \cdot \square \cdot \square$$

Divide the numerator and denominator by common factors.

$$\frac{24}{36} = \frac{\square \cdot \square \cdot \square \cdot \square}{\square \cdot \square \cdot \square \cdot \square}$$

Multiply the remaining factors.

Answer:

Example	Notes
3. Write $\dfrac{150}{240}$ in simplest form using the prime factors method.	
Answer:	
4. Write $\dfrac{21x^2}{24x}$ in simplest form using the prime factors method.	
Answer:	

Helpful Hints

When simplifying fractions using the prime factor method, you must make sure that both the numerator and the denominator are written as the product of only prime numbers and variables without exponents.

If the numerator is a factor of the denominator, then the simplified fraction will have a numerator of 1.

Concept Check

1. How is the prime factors method for simplifying fractions similar to the GCF method?

Practice

Write in simplest form using the prime factors method.

2. $\dfrac{24}{27}$

3. $\dfrac{35}{175}$

4. $\dfrac{-160x}{192xy}$

5. $\dfrac{180y}{210y^2}$

Factors and Fractions
Topic 5.6 Multiplying Fractions

Vocabulary
numerator   •   denominator   •   multiplying fractions

1. Multiplying fractions can be done by first multiplying the numerators to get the
   _____ of the product and multiplying the denominators to get the
   _____ of the product.

Step-by-Step Video Notes
Watch the Step-by-Step Video lesson and complete the examples below.

Example	**Notes**
2. Multiply. Write your answer in simplest form. $$\frac{3}{8} \cdot \frac{7}{4}$$ Multiply the numerators to get the numerator of the product. $$\frac{3}{8} \cdot \frac{7}{4} = \frac{\Box}{?}$$ Multiply the denominators to get the denominator of the product. Simplify. Answer:	
4. Multiply. Write your answer in simplest form. $$\frac{7x}{10} \cdot \frac{5}{8x}$$ Answer:	

69

Example	Notes
7. Multiply $-3x^2 \cdot \dfrac{2}{9x}$ by simplifying first. Write $-3x^2$ as a fraction. Write the problem as one fraction. Answer:	
8. Elizabeth bought a sandwich and ate half of it. She gave her brother half of what was left. How much of the sandwich did her brother get? Answer:	

Helpful Hints

When the products of the numerators and denominators are large, this can make the resulting fraction difficult to simplify. To avoid this, divide the numerator and the denominator by any common factors before multiplying.

Concept Check

1. Which product, $\dfrac{7}{100} \cdot \dfrac{10}{21}$ or $\dfrac{5}{16} \cdot \dfrac{7}{11}$, will need to be simplified? Explain.

Practice

Multiply. Write your answer in simplest form.

2. $\dfrac{8}{15} \cdot \dfrac{3}{5}$

3. $-\dfrac{2}{9} \cdot \dfrac{3}{5}$

4. $\dfrac{7x^2}{9} \cdot \left(-\dfrac{3}{28x}\right)$

5. Lee ate three fourths of a pie, and Kara ate half as much as Lee. How much of the pie did Kara eat?

Factors and Fractions
Topic 5.7 Dividing Fractions

Vocabulary
reciprocals   •   common factors   •   opposites

1. Two numbers are _____ of each other if their product is 1.

Step-by-Step Video Notes
Watch the Step-by-Step Video lesson and complete the examples below.

Example	Notes
1 & 2. Find the reciprocal of the following terms. $\dfrac{3}{4}$ -5	
5. Divide $\dfrac{3}{5} \div \dfrac{1}{2}$. Write your answer in simplest form. Invert the second fraction, and multiply it by the first fraction. $\dfrac{3}{5} \div \dfrac{1}{2} = \dfrac{3}{5} \cdot \dfrac{\square}{\square}$ Rewrite the problem using one fraction. Divide by common factors. Multiply the remaining factors. Answer:	

Example	Notes
7. Divide $\frac{6x}{7} \div \frac{3x}{2}$. Write your answer in simplest form. Answer:	
8. Jeff needs to cut 6 yards of fabric into pieces that are each $\frac{2}{3}$ yards long. How many pieces of cloth can Jeff make? Answer:	

Helpful Hints

To find the reciprocal of an integer term, first write it as a fraction by putting it over 1.

Do not divide by common factors before you invert and multiply.

Because 0 has no reciprocal, any algebraic expression equivalent to zero has no reciprocal. When finding the reciprocal, we assume all algebraic expressions are not equal to zero.

Concept Check

1. William divided $\frac{9}{13} \div \frac{3}{4}$ and got an answer of $\frac{27}{52}$. Find and correct his error.

Practice

2. Divide $\frac{8}{15} \div \frac{3}{4}$. Write your answer in simplest form.

3. Divide $8 \div \frac{2}{9}$. Write your answer in simplest form.

4. Divide $\frac{2}{7} \div 6$. Write your answer in simplest form.

5. Laura needs to separate 6 pounds of hamburger into patties that each weigh $\frac{2}{5}$ pounds. How many patties can she make?

LCM and Fractions
Topic 6.1 Finding the LCM - List Method

Vocabulary
multiple   •   common multiple   •   least common multiple (LCM)   •   whole number

1. A _____ of a term is the product of the term and a positive whole number.

2. The _____ of two or more numbers is the smallest number that is a multiple of the given numbers.

Step-by-Step Video Notes
Watch the Step-by-Step Video lesson and complete the examples below.

Example	Notes

1. Find the first five multiples of 9.
 	
    9, 18, ☐, ☐, ☐
 	
    Answer:	

2. Find the first four multiples of 20*r*.
 	
    20*r*, ☐, ☐, ☐
 	
    Answer:	

3. Find the least common multiple of 9 and 12.
 	
    List the first several multiples of 9.
 	
    List the first several multiples of 12.
 	
    Identify multiples common to both lists.
 	
    Choose the least of these numbers.
 	
    Answer:	

Example	Notes
5. Find the least common multiple of 6, 8, and 12.	
Answer:	
6. Find the least common multiple of $10n$ and $15n$.	
Answer:	

Helpful Hints

Be careful not to confuse the LCM, the least common multiple, with the GCF, the greatest common factor. The LCM is the smallest number that is a multiple of the given numbers, while the GCF is the greatest factor that can be divided evenly into the given numbers.

If one number divides exactly into the other, the LCM is the larger of the two numbers.

Concept Check

1. Latisha states that the LCM of 4 and 10 is 40 because $4 \cdot 10 = 40$. Is she correct? Explain.

Practice

2. List the first five multiples of 11.

3. List the first five multiples of 15.

4. Find the least common multiple of $18y$ and $45y$.

5. Find the least common multiple of 7, 9, and 21.

LCM and Fractions
Topic 6.2 Finding the LCM - GCF Method

Vocabulary
multiple • greatest common factor (GCF) • least common multiple (LCM) • factor

1. If the _____ of two numbers is equal to 1, then the _____ is the product of the two numbers.

Step-by-Step Video Notes
Watch the Step-by-Step Video lesson and complete the examples below.

Example	Notes
1. Find the LCM of 6 and 9 using the GCF method. Find the GCF of 6 and 9. $GCF = \square$ Multiply 6 and 9. $6(9) = \square$ Divide this product by the GCF. $\dfrac{\square}{\square} = \square$ Answer:	
2. Find the LCM of 6 and 15 using the GCF method. Find the GCF of 6 and 15. Multiply 6 and 15. Divide this product by the GCF. Answer:	

Example	Notes
3. Find the LCM of 4 and 18 using the GCF method. Answer:	
4. Find the LCM of 9 and 16 using the GCF method. Answer:	

Helpful Hints

To find the LCM of two numbers using the GCF method, first find the GCF of the two numbers. Then, multiply the two original numbers. Finally, divide this product by the GCF.

Finding the LCM of two or more numbers using the GCF method can be quicker than listing out multiples of each number.

Concept Check

1. Explain why using the GCF method to find the LCM of 3 and 42 can be a better strategy than listing the factors.

Practice

Find the LCM using the GCF method.

2. 6 and 54

4. 8 and 20

3. 10 and 25

5. 5 and 11

LCM and Fractions
Topic 6.3 Finding the LCM - Prime Factor Method

Vocabulary

prime factor method   •   common prime factors   •   prime factorization   •   multiples

1. The _____ of any whole number is the factored form in which all factors are prime numbers.

Step-by-Step Video Notes

Watch the Step-by-Step Video lesson and complete the examples below.

Example	Notes
1 & 2. Find the prime factorization. $12 = \square \cdot \square \cdot \square$ 180	
3. Find the LCM of 12 and 42. Find the prime factorization of each number. 12 42 Write the prime factorizations one below the other, putting prime factors common to both below each other, in columns, when possible. Write down the prime factor from each column. $12 = \square \cdot \square \cdot \square$ $42 = \square \;\downarrow\; \cdot \square \cdot \square$ $\quad \downarrow \;\; \downarrow \;\; \downarrow \;\; \downarrow$ $\text{LCM} = \square \cdot \square \cdot \square \cdot \square$ Multiply the list of prime factors to find the least common multiple. Answer:	

Example	Notes
5. Find the LCM of 6, 18, and 24. Answer:	
7. Find the LCM of $13s^2$ and $19r$. Answer:	

Helpful Hints

When using the prime factor method to find the LCM, make sure that each prime factor of both original numbers is in the LCM.

To find the LCM of two variable factors, choose the one with the highest exponent.

The LCM of any list of distinct prime numbers is the product of those distinct prime numbers. Note that 0 and 1 are not prime numbers.

Concept Check

1. How many total prime factors (matching the total number of columns) are there when finding the LCM of 24 and 30?

Practice

Find the LCM.

2. 8 and 27

3. 4, 5, and 20

4. $6x^2$ and $70x^2 y$

5. $2y^3$, $9r^5$, and $15r^2$

LCM and Fractions
Topic 6.4 Writing Fractions with an LCD

Vocabulary
least common denominator (LCD)   •   common factor   •   equivalent fractions

1. The _____ is the least common multiple (LCM) of the denominators.

2. Fractions are _____ if they have the same value.

Step-by-Step Video Notes
Watch the Step-by-Step Video lesson and complete the examples below.

Example	**Notes**
1. Find the LCM of 4 and 12. Answer:	
2. Find the LCD of $\dfrac{3}{4}$ and $\dfrac{5}{12}$. Answer:	
3. Write a fraction equivalent to $\dfrac{1}{5}$ using the denominator of 15. $\dfrac{1}{5} = \dfrac{\square}{15}$ Answer:	
4. Rewrite $\dfrac{1}{6}$ and $\dfrac{7}{9}$ using the LCD as the denominator. LCD = LCM of $\square$ and $\square = \square$ Answer:	

Example	Notes
6. Rewrite $\dfrac{3}{4x}$ and $\dfrac{1}{2x}$ using the LCD as the denominator. Answer:	

Helpful Hints

If one denominator divides exactly into the other, then the LCD is the larger number.

If two denominators have no common factor other than 1, then the LCD is the product of the two denominators.

The exact same process is used to find both the LCM and the LCD, since finding the LCD is the same as finding the LCM of the denominators.

Concept Check

1. When rewriting $\dfrac{1}{3}$ and $\dfrac{1}{5}$ using the LCD, how do you know that neither numerator will be 1?

Practice

2. Write a fraction equivalent to $\dfrac{2}{3}$ using the denominator of 45.

3. Write a fraction equivalent to $\dfrac{3}{14}$ using the denominator of 98.

4. Rewrite $\dfrac{2}{5}$, $\dfrac{7}{10}$, and $\dfrac{5}{6}$ using the LCD as the denominator.

5. Rewrite $\dfrac{4}{21r}$, $\dfrac{7}{9r^4}$, and $\dfrac{19}{98r^2}$ using the LCD as the denominator.

LCM and Fractions
Topic 6.5 Adding and Subtracting Like Fractions

Vocabulary
numerators   •   like fractions   •   equivalent fractions   •   unlike fractions

1. Fractions with the same, or common, denominator are called _____.

2. Fractions without common denominators are called _____.

Step-by-Step Video Notes
Watch the Step-by-Step Video lesson and complete the examples below.

Example	Notes
1 & 2. Add the following like fractions. Simplify if possible. $$\frac{2}{5}+\frac{1}{5}=\frac{\square+\square}{\square}$$ $$\frac{3w}{10}+\frac{3w}{10}$$	
3 & 4. Subtract the following like fractions. Simplify if possible. $$\frac{11}{14}-\left(-\frac{5}{14}\right)=\frac{11}{14}-\left(\frac{\square}{14}\right)=\frac{\square-(\square)}{\square}$$ $$\frac{6}{7h}-\frac{2}{7h}$$	

Example	Notes
6. Add the fractions $\dfrac{4}{11}+\dfrac{3}{11}+\left(-\dfrac{5}{11}\right)$. Simplify if possible.	
Answer:	
8. Shauna ate $\dfrac{1}{8}$ of a pizza and Kenzi ate $\dfrac{3}{8}$ of the same pizza. Together, how much of the pizza did the two girls eat? 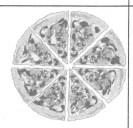	
Answer:	

Helpful Hints

When adding or subtracting like fractions, the operation is done on the numerators only. The denominator remains unchanged.

Concept Check

1. Whether adding or subtracting $\dfrac{4}{5}$ and $\dfrac{2}{5}$, what will stay the same in the result?

Practice

Add. Simplify if possible.

2. $\dfrac{4}{11}+\dfrac{3}{11}$

3. $\dfrac{1}{18}+\dfrac{2}{18}+\dfrac{11}{18}$

Add. Simplify if possible.

4. $\dfrac{5}{9y}+\left(-\dfrac{1}{9y}\right)$

5. Sarah ate $\dfrac{1}{12}$ of a candy bar. Kevin ate another $\dfrac{7}{12}$ of the same candy bar. Together, how much of the candy bar did they eat?

LCM and Fractions
Topic 6.6 Adding and Subtracting Unlike Fractions

Vocabulary
like fractions • unlike fractions • least common denominator (LCD) • denominator

1. Fractions without a common denominator are called _____.

Step-by-Step Video Notes
Watch the Step-by-Step Video lesson and complete the examples below.

Example	Notes
2. Add $\dfrac{3}{4}+\dfrac{1}{6}$. Simplify if possible.	
Find the LCD.	
Rewrite $\dfrac{3}{4}$ with the LCD as the denominator.	
Rewrite $\dfrac{1}{6}$ with the LCD as the denominator.	
Add the numerators. Keep the denominator.	
$\dfrac{3}{4}+\dfrac{1}{6}=\dfrac{\square}{\square}+\dfrac{\square}{\square}=\dfrac{\square}{\square}$	
Answer:	
3. Add $-\dfrac{2}{5}+\left(-\dfrac{3}{8}\right)$. Simplify if possible.	
Answer:	

Example	Notes
6. Subtract $-\dfrac{7}{10x}-\dfrac{2}{15x}$. Simplify if possible.	
Answer:	

7. Paul ate $\dfrac{3}{8}$ of an apple pie.

His brother Jeff ate $\dfrac{1}{4}$ of the same pie. How much more pie did Paul eat than Jeff?

Answer:

Helpful Hints

Be sure you have a common denominator when adding or subtracting fractions.

When rewriting fractions with a common denominator, the LCD is usually used.

Concept Check

1. What important step(s) must be done before adding or subtracting unlike fractions?

Practice

Add or subtract. Simplify if possible.

2. $\dfrac{11}{12}-\dfrac{7}{8}$

3. $\dfrac{5}{6}+\dfrac{7}{10x}$

4. $\dfrac{9}{14a}-\dfrac{13}{15a}$

5. Kevin is planning to cook two recipes. One recipe calls for $\dfrac{1}{4}$ cup of milk, while the other calls for $\dfrac{2}{3}$ cup of milk. How much milk does Kevin need?

Mixed Numbers
Topic 7.1 Changing a Mixed Number to an Improper Fraction

Vocabulary
mixed number   •   improper fraction   •   numerator   •   whole number

1. A(n) _____ is the sum of a whole number and a fraction.

Step-by-Step Video Notes
Watch the Step-by-Step Video lesson and complete the examples below.

Example	Notes
1. Write $2\frac{1}{3}$ as an improper fraction. Multiply the denominator by the whole number. Add the numerator to this product. Write this value over the original denominator. Answer:	
2. Write $4\frac{5}{7}$ as an improper fraction. Answer:	
3. Write $-6\frac{8}{9}$ as an improper fraction. Answer:	

Example	Notes
4. Let's look at a real world example of changing whole numbers to improper fractions. Exchange a $5 bill for quarters at a car wash. Answer:	
5. Write 91 as an improper fraction. Answer:	

Helpful Hints

When you see a negative sign in front of a mixed number, the negative sign applies to both the whole number and the fraction.

To write a negative mixed number as a fraction, treat the number like it is positive and add the negative sign at the end.

Concept Check

1. Show that $2\dfrac{4}{5}$ and $\dfrac{42}{15}$ are equivalent.

Practice

Write each as an improper fraction.

2. $5\dfrac{9}{20}$

4. 82

3. $10\dfrac{4}{7}$

5. $12\dfrac{1}{6}$

Mixed Numbers
Topic 7.2 Changing an Improper Fraction to a Mixed Number

Vocabulary
remainder   •   quotient   •   numerator   •   denominator

1. When changing an improper fraction to a mixed number, the _____ is the whole number part.

Step-by-Step Video Notes
Watch the Step-by-Step Video lesson and complete the examples below.

Example	**Notes**
1. Write $\dfrac{13}{5}$ as a division problem. $13 \div \square$ Answer:	
2. Write $\dfrac{13}{4}$ as a mixed number. Divide the numerator by the denominator. $\overset{\square}{4\overline{)13}}$ The quotient is the whole number part. The remainder is the numerator of the fractional part. The denominator stays the same. $\square\dfrac{\square}{4}$ Simplify the fractional part, if possible. Answer:	

Example	Notes
4 & 5. Write each as a mixed number or an integer. $-\dfrac{15}{5}$ $\dfrac{35}{15}$	
7. Write $-\dfrac{14}{9}$ as a mixed number or an integer. Answer:	

Helpful Hints

Be careful when setting up the division problem. The denominator goes on the outside of the symbol and the numerator goes on the inside, $\dfrac{\text{numerator}}{\text{denominator}} = \text{denominator}\overline{)\text{numerator}}$.

If the remainder is zero when changing an improper fraction into a mixed number, then the answer is an integer.

Concept Check

1. When changing an improper fraction to a mixed number, what does the numerator of the fractional part represent?

Practice

Write each as a mixed number or integer.

2. $\dfrac{11}{3}$

4. $\dfrac{42}{6}$

3. $-\dfrac{17}{4}$

5. $-\dfrac{51}{9}$

Mixed Numbers
Topic 7.3 Multiplying Mixed Numbers

Vocabulary
factor   •   mixed number   •   whole number   •   improper fraction

1. When multiplying mixed numbers, the first step is to rewrite each mixed number as a(n) _____.

Step-by-Step Video Notes
Watch the Step-by-Step Video lesson and complete the examples below.

Example	Notes
2. Multiply $1\frac{2}{3} \cdot 4\frac{1}{5}$. Write each mixed number as an improper fraction. Multiply the fractions. $$\frac{\square}{3} \cdot \frac{\square}{5} = \frac{\square}{\square}$$ Rewrite the product as a mixed or whole number. Answer:	
4. Multiply $2\left(-2\frac{1}{3}\right)\left(-1\frac{3}{4}\right)$. Write each mixed number as an improper fraction. Answer:	

Example	Notes
5. Ethan ran $2\frac{1}{4}$ miles each day for 3 days. How far did Ethan run during the 3-day period?	
Understand the problem and create a plan.	
Find the answer.	
Check the answer.	
Answer:	

Helpful Hints

Remember when multiplying fractions, the numerators are multiplied and the denominators are multiplied.

Make sure to simplify the fraction answer in either the resulting improper fraction or the fraction of the mixed number.

Concept Check

1. Saul thinks the product of $2\frac{3}{8}$ and $6\frac{3}{4}$ is $12\frac{9}{32}$. Explain his error.

Practice

Multiply.

2. $2\frac{2}{3} \cdot 1\frac{1}{4}$

3. $1\frac{2}{7} \cdot \left(-2\frac{1}{3}\right)$

4. $-1\frac{5}{6} \cdot 2\frac{4}{7}$

5. Kevin used $3\frac{1}{2}$ bags of cement to make a staircase. Each bag weighed $4\frac{2}{5}$ pounds. How many pounds of cement did Kevin use?

Mixed Numbers
Topic 7.4 Dividing Mixed Numbers

Vocabulary
mixed number　•　whole number　•　factor　•　improper fraction

1. When dividing mixed numbers, each mixed number should be written as a(n)
    _____.

Step-by-Step Video Notes
Watch the Step-by-Step Video lesson and complete the examples below.

Example	Notes
1. Divide $1\frac{3}{4} \div 2\frac{4}{5}$.	

Write each mixed number as an improper fraction.

$$1\frac{3}{4} \div 2\frac{4}{5} = \frac{\square}{4} \div \frac{\square}{5}$$

Divide the fractions by inverting the second fraction and multiplying it by the first fraction.

If the answer is an improper fraction, rewrite it as a mixed number or whole number.

Answer:

3. Divide $-4 \div 3\frac{1}{3}$.

$$-4 \div 3\frac{1}{3} = -\frac{\square}{\square} \div \frac{\square}{3}$$

Answer:

Example	Notes
5. Megan purchased $12\frac{1}{4}$ yards of fabric. How many dresses can she make if it takes $2\frac{1}{3}$ yards of fabric to make each dress? Answer:	

Helpful Hints

Remember that whole numbers can be rewritten as improper fractions by placing the original number as the numerator and making the denominator 1.

Always simplify a fraction by dividing both numerator and denominator by common factors.

Concept Check

1. Matthias divided $4\frac{2}{3}$ by $2\frac{5}{6}$ and got $2\frac{4}{5}$. Explain and correct his error.

Practice

2. Divide $-3\frac{5}{6} \div \frac{1}{6}$.

4. Divide $6\frac{9}{10} \div 2\frac{4}{5}$.

3. Divide $4\frac{2}{5} \div \left(-2\frac{3}{4}\right)$.

5. A recipe requires $1\frac{2}{3}$ cup of milk per batch. If Joshua has 1 gallon (16 cups) of milk, how many full batches can Joshua make?

Mixed Numbers
Topic 7.5 Adding Mixed Numbers

Vocabulary
mixed number   •   common denominator   •   numerator   •   denominator

1. When adding mixed numbers, if necessary, rewrite the fractions as equivalent fractions that have a _____.

Step-by-Step Video Notes
Watch the Step-by-Step Video lesson and complete the examples below.

Example	**Notes**
2. Add $7\frac{1}{3} + 4\frac{3}{5}$. Rewrite the fractions as equivalent fractions with the LCD. $7\frac{1}{3} + 4\frac{3}{5} = 7\frac{\square}{\square} + 4\frac{\square}{\square}$ Add the fractions. Add the integers. If the fraction is improper, change to a mixed number and add the integers. Simplify if possible. Answer:	
4. Add $-2\frac{1}{4} + 15\frac{2}{3}$. Answer:	

Example	Notes
5 & 6. Add. $2 + 4\dfrac{2}{5}$ $2\dfrac{5}{6} + \left(-\dfrac{3}{5}\right)$	

7. Jordan bought $4\dfrac{2}{3}$ pounds of ham and $1\dfrac{1}{2}$ pounds of roast beef. How many pounds of lunch meat did Jordan buy?

     Answer:

Helpful Hints

If one of the mixed numbers is negative, treat both the integer and the fraction like they are negative.

Concept Check

1. Marsha added $4\dfrac{5}{7}$ and $1\dfrac{3}{7}$ on a test and got $5\dfrac{8}{7}$. She checked her work and it was correct, but she still lost points for the answer. Explain and correct her error.

Practice

2. Add $-4\dfrac{1}{5} + 2\dfrac{2}{5}$.

3. Add $2\dfrac{3}{4} + 3\dfrac{1}{4}$.

4. Add $1\dfrac{7}{10} + \left(-2\dfrac{4}{5}\right)$.

5. Jordan had $1\dfrac{1}{3}$ pounds of roast beef left the next week. He went to the store and bought $2\dfrac{1}{2}$ pounds of turkey and $\dfrac{1}{4}$ pound of ham. How much lunch meat does Jordan have now?

Mixed Numbers
Topic 7.6 Subtracting Mixed Numbers

Vocabulary
mixed number   •   denominator   •   borrowing   •   improper fraction

1. Consider $4\frac{2}{5} - 2\frac{3}{5}$. Since $\frac{2}{5}$ is less than $\frac{3}{5}$, _____ is needed in order to subtract.

Step-by-Step Video Notes
Watch the Step-by-Step Video lesson and complete the examples below.

Example	Notes
2. Subtract $6\frac{1}{2} - 3\frac{1}{4}$.	
Rewrite the fractions as equivalent fractions with the LCD.	
$6\frac{1}{2} - 3\frac{1}{4} = 6\frac{\square}{\square} - 3\frac{\square}{\square}$	
Borrowing _____ needed.	
Subtract the fractions. Subtract the integers. Simplify if possible.	
Answer:	
4. Subtract $-7\frac{4}{5} - 4\frac{3}{5}$.	
Answer:	

Example	Notes
5 & 6. Subtract. $$4 - 1\frac{1}{8}$$ $$8\frac{6}{7} - \left(-1\frac{1}{7}\right)$$	

7. James needs a piece of wood $5\frac{1}{3}$ feet long and buys an 8-foot board. If he cuts a piece $5\frac{1}{3}$ feet long, how long is the remaining piece of wood?

   Answer:

Helpful Hints

When borrowing in a mixed number subtraction, subtract 1 from the integer and add it to the fraction in terms of the denominator. Borrowing is also called regrouping.

When subtracting mixed numbers, if necessary, rewrite the fractions as equivalent fractions that have a common denominator, usually the LCD.

Concept Check

1. Jim subtracted $4\frac{3}{7}$ from $6\frac{4}{9}$ and got $2\frac{1}{2}$ as his answer. Explain and correct his error.

Practice

2. Subtract $-2\frac{4}{5} - 1\frac{2}{5}$.

3. Subtract $5\frac{1}{3} - 2\frac{2}{3}$.

4. Subtract $-5\frac{1}{3} - \left(-2\frac{1}{6}\right)$.

5. Saul has $3\frac{1}{3}$ cups of milk. He needs $2\frac{3}{4}$ cups of milk for a recipe. How much milk will he have left over after he makes the recipe?

Mixed Numbers
Topic 7.7 Adding and Subtracting Mixed Numbers - Improper Fractions

Vocabulary
mixed numbers   •   improper fractions   •   numerators   •   least common denominators

1. When adding and subtracting mixed numbers, the mixed numbers can be rewritten as _____.

Step-by-Step Video Notes
Watch the Step-by-Step Video lesson and complete the examples below.

Example	Notes
1. Add $2\frac{4}{5}+1\frac{2}{5}$. Write each mixed number as an improper fraction. $\dfrac{\Box}{5}+\dfrac{\Box}{5}$ Add the fractions. If the answer is an improper fraction, change it back to a mixed number. Simplify, if possible. Answer:	
2. Subtract $3\frac{4}{5}-2\frac{2}{5}$. Answer:	

Example	Notes
5. Add $1\dfrac{2}{5}+3\dfrac{1}{2}$.	
Answer:	
7. Kenny ran $6\dfrac{3}{10}$ km in 1 hour on the first day of his new workout plan. On the second day, he ran $8\dfrac{1}{8}$ km in 1 hour. How much farther did Kenny run the second day?	
Answer:	

Helpful Hints
Writing mixed numbers as improper fractions before adding allows you to avoid changing improper fractions to mixed numbers and carrying a one to the whole number part.

Concept Check
1. What process does converting the mixed numbers into improper fractions eliminate in subtraction problems?

Practice
Add or subtract.

2. $-2\dfrac{5}{7}+1\dfrac{3}{7}$

3. $4\dfrac{2}{3}-1\dfrac{1}{3}$

4. $3\dfrac{1}{4}-\left(-1\dfrac{1}{2}\right)$

5. Refer to Example 7. What is the total distance Kenny ran on the first two days of his new workout plan?

Name: _____ Date: _____

Instructor: _____ Section: _____

Operations with Decimals
Topic 8.1 Decimal Notation

Vocabulary
place value　　•　　standard form　　•　　decimal point　　•　　decimal fraction

1. A decimal, also known as a decimal number, has three parts: a whole number part, a
_____, and a decimal part.

Step-by-Step Video Notes
Watch the Step-by-Step Video lesson and complete the examples below.

Example	Notes
1–3. Write each decimal in words. 0.32 15.073 2.25	
6. Read the following decimal. 12.0457 Answer:	
7 & 8. Write each decimal in standard form. twenty-one and two hundred thirty-seven thousandths fourteen and eight hundredths	

Example	Notes
9. Write the decimal fraction and the decimal number that represents the shaded part. 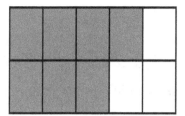 Answer:	

Helpful Hints

When reading a decimal number or writing a decimal number in words, use the word "and" in place of the decimal point.

To write a decimal in standard form, first write the whole number in number form. Then, write the decimal point in place of the word "and." Use the given place value to determine the number of decimal places. Finally, write the decimal part in number form so that it ends at the given place value, inserting zeros at the beginning if needed.

Concept Check
1. How many zeros should you insert after the decimal point when writing the number four and three thousandths in standard form?

Practice

Write each decimal in words.

2. 301.03

3. 4.718

Write each decimal in standard form.

4. twenty-seven thousandths

5. five hundred ten and nine tenths

Operations with Decimals
Topic 8.2 Comparing Decimals

Vocabulary
inequality symbols   •   is less than   •   is greater than   •   comparing decimals

1. The symbol < means _____.

2. The symbol > means _____.

Step-by-Step Video Notes
Watch the Step-by-Step Video lesson and complete the examples below.

Example	**Notes**
1–3. Fill in the blank with < or > to make a true statement. 0.14 ☐ 0.41 2.875 ☐ 2.835 41.13 ☐ 41.92	
4. Which is larger, 0.6 or 0.59? Answer:	
5–7. Fill in the blank with < or > to make a true statement. −0.2 ☐ −0.6 −0.9822 ☐ −0.783 −4.13 ☐ 4.9	

Example	Notes
8. Order the scores from smallest to largest. During a gymnastics event, a women's gymnastics team scored 46.875 on the vault, 47.975 on the uneven bars, 47.25 on the balance beam, and 44.425 on the floor exercises. Answer:	

Helpful Hints

The inequality symbol always points to the smaller number.

Adding zeros to the end of a decimal *does not* change its value. Inserting zeros at the beginning of the decimal part of a number *does* change its value. For example, 0.6 is the same as 0.60, but it is not the same as 0.06.

A positive decimal will always be larger than a negative decimal.

Concept Check

1. What is a way to compare decimal numbers with a different number of decimal places?

Practice

Fill in the blank with $<$ or $>$ to make a true statement.

2. 0.53 ☐ 0.503

3. -6.3 ☐ -6.29

Arrange from smallest to largest.

4. $1.104, \ 1.04, \ 1.4, \ 1.14$

5. $0.3, \ 0.33, \ -3.003, \ 0.033$

Name: _____   Date: _____

Instructor: _____   Section: _____

Operations with Decimals
Topic 8.3 Rounding Decimals

Vocabulary
sales tax • batting average • round up • round down

1. You underline the digit in the place value to which you are rounding. If the digit to the right is 5 or more, you will _____, or increase the underlined digit by 1.

Step-by-Step Video Notes
Watch the Step-by-Step Video lesson and complete the examples below.

Example	Notes
1. Rasheed buys a camera for $328.28. There is a sales tax, which calculates to $20.5175. The store needs to round this tax to the nearest cent, or hundredths place, before adding it to Rasheed's bill.	

Whole Numbers				.	Decimals				
One Thousands	Hundreds	Tens	Ones	Decimal Point	Tenths	Hundredths	One – Thousandths	Ten – Thousandths	Hundred – Thousandths

Underline the digit in the place value to which you are rounding.

20.5<u>1</u>75

Look at the digit to the right of the underlined digit. If that digit is 4 or less, round down; if it is 5 or more, round up. That digit is ☐, so round up to 2. Leave off all digits to the right of the underlined digit.

Answer:

Example	Notes
5. Round 13.45812 to the nearest thousandth. Underline the digit in the place value to which you are rounding. 13.45812 Look at the digit to the right of the underlined digit. Round up or round down. Answer:	
7. Round 7.9561 to the nearest tenth. Answer:	

Helpful Hints

Many common statistics, for example grade point average and sports statistics are rounded to a specific place value. Many common money transactions like interest, taxes, and discounts are rounded to the nearest cent.

If the digit you are rounding up is a 9, you change that digit to a 0, and increase the digit to the left by 1. For example, 4.972 rounded to the nearest tenth is 5.0

Leave off all digits to the right of the place to which you are rounding. If you round a decimal to the nearest whole number, do not write a decimal point or any decimal places.

Concept Check
1. Which number has more decimal places, 5.325 rounded to the nearest hundredth, or 6.357895 rounded to the nearest tenth?

Practice

Round to the nearest hundredth.
2. 62.514

3. 99.999

Round to the nearest tenth.
4. 34.96

5. 0.285714

Operations with Decimals
Topic 8.4 Adding and Subtracting Decimals

Vocabulary
estimation • decimal places • decimal points • perimeter of a triangle

1. When adding or subtracting decimals, write the numbers vertically, making sure the
   _____ line up.

Step-by-Step Video Notes
Watch the Step-by-Step Video lesson and complete the examples below.

Example	Notes
1. Jennifer spent $2.97 for a meal and $1.19 for dessert, including tax. How much did she spend in total? Write the numbers vertically, so that the digits with the same place value are in the same column, and the decimal points are lined up as well. 2. 9 7 + 1. 1 9 ☐.☐☐ Answer:	
3 & 4 Find the sum or difference. $-14.74 + 6.13$ $4.09 - (-5.85)$	

Example	Notes
5. Find the difference. $13.24f - 7f$ Answer:	
8. At Sandee's Malt Shop, a hamburger costs $3.95, a hot dog costs $2.49, fries cost $1.59, and onion rings cost $1.89. How much more does a hamburger with fries cost than a hot dog with onion rings? Answer:	

Helpful Hints

Adding and subtracting decimals is similar to adding and subtracting whole numbers and like fractions. Add or subtract "like" parts; ones to ones, tens to tens, tenths to tenths, hundredths to hundredths, etc.

Adding and subtracting negative decimals follows the same procedure as adding and subtracting integers.

Concept Check

1. To subtract $24.3 - 7.52$ vertically, which number will need a zero as a placeholder?

Practice

Find the sum or difference.

2. $5.27 + 12.38$

3. $17.9x - 8.46x$

4. $4.17 + 23.95 - 17.25$

5. At a ski shop, leather gloves cost $42.45, a leather coat costs $199.89, an insulated coat costs $240.42, and insulated gloves cost $19.51. How much more expensive is an insulated coat and insulated gloves than a leather coat and leather gloves?

Operations with Decimals
Topic 8.5 Multiplying Decimals

Vocabulary
power of 10   •   factor   •   decimal   •   whole number

1. To multiply a decimal by a _____, move the decimal point to the right the same number of places as the number of zeros in the power of 10.

Step-by-Step Video Notes
Watch the Step-by-Step Video lesson and complete the examples below.

Example	**Notes**
1. Multiply 3.4×6.1.	

$$\begin{array}{r} 3.\ 4 \\ \times\ 6.\ 1 \\ \hline 3\ 4 \\ \square\ \square\ \square\ 0 \\ \hline \square\ \square\ \square\ 4 \end{array}$$

Answer:

2 & 3. Multiply.

$-4(2.5)$

0.152×0.23

4. $2.7f \times 0.64$

Answer:

Example	Notes
5. Multiply 5.793×100. Move the decimal point ☐ places to the right. Answer:	
7. Ian's new car travels 25.6 miles for each gallon of gas. How many miles can he travel on a full tank, which contains 12 gallons of gas? Answer:	

Helpful Hints

After multiplying, insert zeros before the answer if the number of decimal places needed is greater than the number of digits in the answer.

To multiply expressions that have decimal number coefficients, multiply the decimal numbers, then multiply the result by the variable factor(s).

Concept Check

1. Will you need to add zeros at the end of your answer when multiplying 24.043×1000?

Practice

2. Multiply 3.2×5.08.

3. Multiply 7.4×0.006.

4. Multiply $8.13 \times 10,000$.

5. Saul reads that a 12-ounce can of soda contains 40.5 grams of sugar. How many grams of sugar are in a 24-can case of soda?

Operations with Decimals
Topic 8.6 Dividing Decimals

Vocabulary
divisor   •   product   •   dividend   •   quotient

1. The _____ is the answer to the division problem.

Step-by-Step Video Notes
Watch the Step-by-Step Video lesson and complete the examples below.

Example	Notes
1. $6.3 \div 3$ $3\overline{)6.3}$ Answer:	
3. Divide 6.1 by 3. $3\overline{)6.1000}$ Answer:	
5. Divide $2.9 \div 1000$. Answer:	

Example	Notes
6. Divide $0.0612a \div 0.12$. Answer:	
7. Divide $1.219 \div 0.3$. Answer:	
8. Simon, Penny, Ashley, and James went out to eat. The total bill was $36.24. If they split the bill evenly, how much will each person pay? Answer:	

Helpful Hints

To divide a decimal by a whole number, copy the decimal point straight up into the quotient.

When dividing by a decimal, the decimal point is moved the same number of places in the dividend as needed to achieve a whole number in the divisor.

When writing a repeating decimal, draw the bar only over the decimals that repeat.

Concept Check
1. What different steps are taken when dividing a decimal by a decimal vs. dividing it by a whole number?

Practice
Divide.

2. $8.5 \div 5$

3. $5.3x \div 4$

4. $9.4 \div 100$

5. Three roommates equally share the cost of buying a new TV. If the TV costs $618.27, how much does each roommate pay?

More with Fractions and Decimals
Topic 9.1 Order of Operations and Fractions

Vocabulary
order of operations • addition of fractions • multiplication of fractions • LCM

1. The procedure for the _____ is to evaluate what is inside parentheses, evaluate any exponents, perform multiplication/division, and then perform addition/subtraction.

Step-by-Step Video Notes
Watch the Step-by-Step Video lesson and complete the examples below.

Example	**Notes**
1. Simplify $\left(\dfrac{1}{3} - \dfrac{1}{6} \right) + \dfrac{1}{2}$. P: Parentheses $\dfrac{1}{3} - \dfrac{1}{6} = \dfrac{\square}{\square}$ A<u>S</u> : Add and Subtract $\dfrac{\square}{\square} + \dfrac{1}{2} = \dfrac{\square}{\square}$ Answer:	
2. Simplify $\dfrac{1}{4} \div \dfrac{3}{2} + \dfrac{1}{2} \cdot \dfrac{1}{3}$. M<u>D</u> : Multiply and Divide A<u>S</u> : Add and Subtract Answer:	

Example	Notes
3. Simplify $\left\| -\dfrac{2}{5}+\dfrac{1}{5} \right\|^{2} \div \dfrac{3}{10}$.	
Answer:	
4. Simplify $\left[2\left(\dfrac{1}{3}\cdot\dfrac{3}{4} \right) \right]^{2} -\dfrac{1}{7}$.	
Answer:	

Helpful Hints

Use the phrase "**P**lease **E**xcuse **M**y **D**ear **A**unt **S**ally" to remember the order of operations, P E MD AS.

After addressing any parentheses and exponents, make sure to do the multiplication/division from left to right, then the addition/subtraction from left to right.

Concept Check

1. What is the second step when simplifying $\dfrac{1}{5}-\dfrac{2}{5}\cdot\left(\dfrac{1}{4} \right)^{2} +\dfrac{5}{6}$?

Practice
Simplify.

2. $\dfrac{1}{7}+\dfrac{2}{7}\cdot\left(\dfrac{1}{2} \right)^{2}$

4. $\dfrac{1}{4}\cdot\left(\dfrac{3}{5}+\dfrac{1}{10} \right)\div\dfrac{3}{10}$

3. $\dfrac{1}{2}-\dfrac{1}{3}\div\dfrac{5}{12}$

5. $\dfrac{4}{5}+\left(2\left\| -\dfrac{3}{5} \right\| \right)^{2}\cdot\dfrac{25}{36}$

More with Fractions and Decimals
Topic 9.2 Order of Operations and Decimals

Vocabulary
addition and multiplication   •   addition and subtraction   •   multiplication and division

1. In the order or operations, after evaluating what is inside the parentheses and evaluating any exponents, the next step is to perform _____ from left to right.

Step-by-Step Video Notes
Watch the Step-by-Step Video lesson and complete the examples below.

Example	Notes
1 & 2. Simplify.	

$3.6 + (0.2 + 0.3)^2$

$3.6 + (\boxed{})^2$

$3.6 + \boxed{} = \boxed{}$

$(12.6 - 11.16) \div 0.2 \times (-3)$

$(\boxed{}) \div 0.2 \times (-3)$

3. Simplify $10.5 - 3.2 + 7.2 \div 3.6$.

Answer:

Example	Notes
5. Simplify $\dfrac{-10.1+2.1(-3)}{5.05+3.15}$.	

Answer: _____

6. Simplify $\left[3-0.2^3\right]-\left[1.1\times(5.8-6.7)\right]$.

Answer: _____

Helpful Hints

The order of operations for decimals is the same as for whole numbers.

Treat a fraction bar like there are parentheses around the numerator and the denominator.

When grouping symbols appear within grouping symbols, simplify the innermost grouping symbols first, and work outward.

Concept Check

1. What is the last operation performed when simplifying $\left[(1+3)^2\cdot3-12\right]\div2$?

Practice
Simplify.

2. $4.1-(0.5+0.4)^2$

3. $1.3+0.5\div(0.9-0.4)$

4. $3+2(0.5)^2$

5. $\dfrac{0.4+4(0.5)}{0.2+0.4}$

More with Fractions and Decimals
Topic 9.3 Converting Fractions to Decimals

Vocabulary
improper fraction   •   comparing decimals   •   repeating decimal   •   equivalent decimal

1. A decimal that repeats in a pattern without end is a(n) _____.

Step-by-Step Video Notes
Watch the Step-by-Step Video lesson and complete the examples below.

Example	Notes
1 & 2. Write each fraction as an equivalent decimal. $$\frac{1}{10}$$ $$\frac{9}{100}$$	
4. Write $\frac{3}{5}$ as an equivalent decimal. $\square\overline{)3.0}$ Answer:	
7. Convert $-\frac{4}{11}$ to a decimal. Round to the nearest thousandth. Answer:	

Example	Notes
8. Convert $2\frac{3}{4}$ to a decimal.	
Answer:	
9. Write this fraction as a decimal rounded to the nearest thousandth. The lead-off batter for the Cleveland Raiders baseball team got on base 31 out of 70 times at bat. Answer:	

Helpful Hints

When converting a fraction to a decimal, divide the numerator by the denominator.

When dividing the numerator by the denominator results in a repeating pattern, placing a bar over the repeating number(s) indicates the repeating decimal.

Concept Check

1. Fred converts $\frac{3}{8}$ to a decimal and incorrectly gets an answer of $2.\overline{6}$. Find and explain his error. Then find the correct decimal equivalent.

Practice

Convert the fraction to a decimal. Use the bar over repeating decimals when necessary.

2. $\frac{42}{100}$

3. $-\frac{13}{25}$

4. $7\frac{1}{6}$

5. Eighteen out of forty-five students passed the test.

More with Fractions and Decimals
Topic 9.4 Converting Decimals to Fractions

Vocabulary
improper fraction   •   denominator   •   numerator   •   simplest form

1. Write the decimal part of the decimal number as the _____ of the fraction.

Step-by-Step Video Notes
Watch the Step-by-Step Video lesson and complete the examples below.

Example	Notes
1. Write 5.277 as a fraction or mixed number in simplest form.	
Write the decimal part of the decimal number as the numerator of the fraction.	
Determine the place value of the last digit in the decimal.	
Keep any whole number as the whole number part of a mixed number.	
Answer:	
2 & 3. Write the decimal as a fraction in simplest form. 0.96 −0.0201	

Example	Notes
4 & 5. Write each decimal as a fraction or mixed number in simplest form. 4.0358 −5.2	
6. Sarah's GPS says that she needs to drive 5.85 miles to get home. Write this distance as a mixed number in simplest form. Answer:	

Helpful Hints

The number of decimal places is equal to the number of zeros in the power of 10 in the denominator. We can also use this to find the denominator.

When converting a decimal to a fraction, the fraction will only simplify if the numerator has a factor of either 2 or 5.

Concept Check

1. What are the numerator and denominator when writing a decimal as a fraction?

Practice

Write each decimal as a fraction in simplest form.

2. 0.303

4. 6.19

3. −1.75

5. Tom bought 3.72 pounds of steak at the butcher. Write this as a mixed number in simplest form.

More with Fractions and Decimals
Topic 9.5 Solving Equations Involving Fractions

Vocabulary
least common denominator (LCD)   •   denominator   •   greatest common factor (GCF)

1. The _____ of two or more fractions is the least common multiple (LCM) of the denominators of the fractions.

Step-by-Step Video Notes
Watch the Step-by-Step Video lesson and complete the examples below.

Example	Notes
1. Solve for x. $\dfrac{1}{4}x - \dfrac{2}{3} = \dfrac{5}{12}x$ Find the LCD of the fractions. Multiply both sides of the equation by the LCD. Use the Distributive Property. Answer:	
2. Solve for x. $\dfrac{x}{3} + 3 = \dfrac{x}{5} - \dfrac{1}{3}$ Answer:	

Example	Notes
3. Solve for x. $$\frac{x}{4} + \frac{1}{2} = \frac{x+5}{7}$$ Answer:	

Helpful Hints

The equation-solving procedures are the same for equations with fractions or without fractions. To make the calculations a little "nicer," we can multiply both sides of the equation by the least common denominator (LCD) of all the fractions contained in the equation. If done correctly, all fractions will change into integers.

Multiplying both sides of an equation by the LCD and then using the Distributive Property is the same as multiplying each term in the equation by the LCD.

Concept Check
1. If both sides of an equation with fractions are multiplied by a number other than the LCD, will it change the solution? Explain.

Practice
Solve for x.

2. $\dfrac{2}{3}x = \dfrac{1}{15}x + \dfrac{3}{5}$

4. $\dfrac{x-5}{4} = 1 - \dfrac{x}{5}$

3. $\dfrac{x}{3} - 1 = -\dfrac{1}{2} - x$

5. $\dfrac{1+2x}{5} + \dfrac{4-x}{3} = \dfrac{1}{15}$

More with Fractions and Decimals
Topic 9.6 Solving Equations Involving Decimals

Vocabulary
least common multiple (LCM)   •   decimal point   •   power of ten   •   decimal

1. The decimal with the most decimal places will determine the _____ that we will multiply each term by.

Step-by-Step Video Notes
Watch the Step-by-Step Video lesson and complete the examples below.

Example	**Notes**
1. Solve for x.	

$$0.1x + 0.5 = 1.2$$

$$\boxed{}(0.1x) + \boxed{}(0.5) = \boxed{}(1.2)$$

$$1x + \boxed{} = \boxed{}$$

Answer: _____

2. Solve for x.

$$0.21x + 4.3 = 3.67$$

$$\boxed{}(0.21x) + \boxed{}(4.3) = \boxed{}(3.67)$$

Answer: _____

Example	Notes
3. Solve for x. $0.5x - 0.3(2 - x) = 4.6$ Answer:	
4. Solve for x. $-0.7b + 1.3 = -9 - 1.5b$ Answer:	

Helpful Hints
If the decimals are tenths, multiply the equation by 10; if the decimals are hundredths, then multiply by 100, etc. If there are numbers with different numbers of decimal places, multiply by a power of ten that will eliminate all the decimal places.

When a problem starts with decimals, the answer should be written in decimal form, instead of fractions or mixed numbers.

Concept Check
1. Lauren wants to solve the equation $0.320x + 4.78 = 3.8$. She starts by multiplying each term by 1000. Is this a correct method of solving this equation? Can you suggest a simpler method for solving the equation? Explain.

Practice
Solve for x.
2. $0.6x + 5.9 = 3.8$

4. $-8(0.1x + 0.4) - 0.9 = -0.1$

3. $-1.2x - 4.3 = 1.1$

5. $0.6(x + 0.1) = 2(0.4x - 0.2)$

Ratios, Rates, and Percents
Topic 10.1 Ratios

Vocabulary

numerator   •   denominator   •   ratio   •   fraction

1. A(n) _____ is a comparison of two like quantities, measured in the same units.

Step-by-Step Video Notes
Watch the Step-by-Step Video lesson and complete the examples below.

Example	Notes
1. A gallon of lemonade is made by mixing 2 cups of lemon juice with 14 cups of water. Write the ratio of lemon juice to water in a gallon of lemonade. Use all three forms. The ratio can be written in words. 2 to 14 The ratio can be written using a colon. 2: ☐ The ratio can be written as a fraction. $\dfrac{2}{14}$ Answer:	
2. Write 5 hours to 7 hours as a ratio. Write ratio in word form. _____ Write the ratio using a colon. ☐ Write the ratio as a fraction. $\dfrac{\square}{\square}$ Answer:	

Example	Notes
4. Write 4 ounces:32 ounces as a fraction in simplest form. $$\frac{4}{\Box} = \frac{\Box}{\Box}$$ Answer:	
5. Write \$1.23 to \$3.75 as a fraction in simplest form. Answer:	

Helpful Hints

Ratios can be written in three ways: in words, using a colon, and as a fraction.

We do not need to include the units in ratios since the units are the same and can be divided out as a common factor.

Concept Check

1. How is a ratio written as a fraction?

Practice

Write the ratio in all three forms.

2. An index card measures 3 inches in width and 5 inches in length. Write the ratio of width to length.

Write the ratio as a fraction in simplest form.

4. 8 inches to 12 inches

3. A classroom has 7 boys and 9 girls. Write the ratio of boys to girls in the class.

5. \$4.25 to \$6.50

Ratios, Rates, and Percents
Topic 10.2 Rates

Vocabulary

rate   •   fraction   •   ratio   •   unit rate

1. A(n) _____ is a ratio that compares two quantities with different units.

2. A(n) _____ gives the rate for one unit of the item.

Step-by-Step Video Notes
Watch the Step-by-Step Video lesson and complete the examples below.

Example	Notes
1. Write the rate of 240 miles per 10 gallons as a fraction is simplest form. The first quantity is written with units as the numerator. Write the second quantity with units as the denominator. $\dfrac{240 \text{ miles}}{\boxed{}\ \underline{}}$ Simplify. $\dfrac{240 \text{ miles}}{\boxed{}\ \underline{}} = \dfrac{\boxed{} \text{ miles}}{\boxed{} \text{ gallon}}$ Answer:	
2. Write the rate 18 tomatoes for 4 pots of stew as a fraction in simplest form. Write the ratio as a fraction include units. $\dfrac{18\ \underline{}}{\boxed{} \text{ pots of stew}}$ Simplify. Answer:	

Example	Notes
4. Write the ratio $250 for 20 hours as a fraction in simplest form. Remember that units are included in a rate. Answer:	
5. A 6-pack of juice bottles costs $3.00. Find the unit rate of cost per 1 bottle. Write the rate as a fraction with units. _____ Perform the division and attach the units. Answer:	

Helpful Hints

When writing a rate as a fraction, remember that the first quantity is the numerator and the second quantity is the denominator.

To find a rate, remember to add the units into the fraction.

Concept Check

1. What is the difference between a rate and a unit rate?

Practice

Write the rate as a fraction in simplest form.

2. 320 miles per 8 hours

Find the unit rate.

4. Carl earns $600 for 40 hours of work

3. 15 gallons of paint to paint 6 apartments

5. Kim spent $450 for 15 gallons of paint.

Ratios, Rates, and Percents
Topic 10.3 Proportions

Vocabulary
ratio   •   equation   •   proportion   •   fraction

1. A(n) _____ is a statement that two quantities are the same, or equal.

2. A(n) _____ is a statement that two rates or ratios are equal.

Step-by-Step Video Notes
Watch the Step-by-Step Video lesson and complete the examples below.

Example	Notes
1. Write the following as a proportion. If 3 inches on a map represent 270 miles, 6 inches represent 540 miles. The first fraction is $\dfrac{3 \text{ inches}}{\boxed{}}$. The second fraction is $\dfrac{\boxed{}}{\boxed{}}$. Set the fractions equal. Answer:	
3. Determine if $\dfrac{6}{21} \overset{?}{=} \dfrac{10}{35}$ is a proportion. Check if the cross products are equal. $6 \cdot 35 = \boxed{}$ $21 \cdot 10 = \boxed{}$ Answer:	

Example	Notes
5. Find the missing number in the proportion. $$\frac{3}{9} = \frac{4}{n}$$ Find the cross products. Answer:	
8. A professional golfer has a score of -3 after a round of 18 holes. If he keeps the same pace in a 72-hole tournament, what would the golfer's final tournament score be? Answer:	

Helpful Hints

When setting up a proportion, units must "match up" or be in the same place in the fractions.

In a proportion, the cross products must be equal.

Concept Check

1. How do you determine if the statement of a fraction equal to a fraction is a proportion?

Practice

2. Write a proportion for 8 is to 12 as 2 is to 3.

3. Is this statement a proportion?
$$\frac{5}{6.2} \overset{?}{=} \frac{6}{7.3}$$

4. Find the missing number in the proportion.
$$\frac{-6}{18} = \frac{5}{n}$$

5. Mark filled his 2-gallon gas can for $7.50. How much will it cost for him to fill his 5-gallon gas can?

Ratios, Rates, and Percents
Topic 10.4 Percent Notation

Vocabulary
rate   •   percent   •   denominator   •   fraction

1. _____ means per 100.

Step-by-Step Video Notes
Watch the Step-by-Step Video lesson and complete the examples below.

Example	Notes
2. Write 14% as a fraction in simplest form. Write the percent number as numerator in a fraction with denominator of 100. $\dfrac{\square}{100}$ Simplify the fraction. $\dfrac{\square}{100} = \dfrac{\square}{\square}$ Answer:	
4. Write $\dfrac{7}{100}$ as a percent. Since the denominator of the fraction is 100, write the numerator as the percent. $\square$ % Answer:	

Example	Notes
9. Is 50% less than, equal to, or greater than one whole unit. 100% represents one whole unit. Answer:	
13. A company produced 2500 personal computers, and 3% of them were defective. How many of the computers were defective? To find 3% of 2500, turn 3% into a fraction. $3\% = \dfrac{\square}{\square}$ Multiply this fraction by 2500. Answer:	

Helpful Hints

To convert p% to a fraction, write $\dfrac{p}{100}$.

Remember that 100% represents one whole unit.

Concept Check
1. How is the percent of a number found?

Practice

2. Write 128% as a fraction in simplest form.

3. Write $\dfrac{29}{100}$ as a percent.

4. Is 225% less than, equal to, or greater than one whole unit?

5. A company has 600 employees, and 52% of them are male. How many employees are male?

Ratios, Rates, and Percents
Topic 10.5 Percent and Decimal Conversions

Vocabulary

decimal   •   percent   •   ratio   •   fraction

1. _____ means per 100.

Step-by-Step Video Notes

Watch the Step-by-Step Video lesson and complete the examples below.

Example	Notes
1. Write 51% as a decimal. Place the decimal at the end of the number and then move it two places to the left. [.] The percent sign is left off. Answer:	
5. Write 4.5% as a decimal. Answer:	
6. Write 0.71 as a percent. Move the decimal two places to the right, adding zeros if needed. 0.71 → [ ] Write the percent sign. Answer:	

Example	Notes
9. Write 0.75 as a percent. Answer:	

Helpful Hints

When converting a percent to a decimal, it may be necessary to add zeros.

When converting a decimal to a percent, it may be necessary to add zeros.

Concept Check

1. What are the similarities and differences in converting a percent to a decimal and converting a decimal to a percent?

Practice

Write the percent as a decimal.

2. 32.5%

Write the decimal as a percent.

4. 1.8

3. 6.1%

5. 0.025

Ratios, Rates, and Percents
Topic 10.6 Percent and Fraction Conversions

Vocabulary

decimal   •   percent   •   ratio   •   fraction

1. If the denominator of a fraction is 100, then the numerator is the _____.

Step-by-Step Video Notes
Watch the Step-by-Step Video lesson and complete the examples below.

Example	Notes
1. Write 8% as a fraction in simplest form. Write the percent as the numerator with the denominator as 100. $\dfrac{\square}{100}$ Simplify. Answer:	
4. The sales tax is 4.5% of the price. Write the percent as a fraction in simplest form. Write the percent as an equivalent fraction. $4.5 = \dfrac{\square}{10}$ Multiply this fraction by $\dfrac{1}{100}$. $\dfrac{\square}{10} \cdot \dfrac{1}{100} = \dfrac{\square}{\square}$ Answer:	

Example	Notes
6. Write $\dfrac{5}{8}$ as a percent. Divide 5 by 8. Convert to a percent by moving the decimal place two places to the right and then write the percent sign. Answer:	
7. Write $\dfrac{4}{7}$ as a percent. Round to the nearest tenth of a percent. Answer:	

Helpful Hints

You can convert p% to a fraction by dividing p by 100 or by multiplying p by $\dfrac{1}{100}$.

Remember that when a fraction has a denominator of 100, the numerator is the percent.

Concept Check

1. What are the similarities and differences in converting a percent to a fraction and converting a fraction to a percent?

Practice

Write the % as a fraction in simplest form.

2. 12%

3. 6.1%

Write the fraction as a decimal. Round to the nearest tenth of a percent as needed.

4. $\dfrac{3}{5}$

5. $\dfrac{5}{6}$

Ratios, Rates, and Percents
Topic 10.7 The Percent Equation

Vocabulary
decimal   •   percent equation   •   base   •   fraction

1. The _____ states "the amount is equal to the percent times the base."

Step-by-Step Video Notes
Watch the Step-by-Step Video lesson and complete the examples below.

Example	**Notes**
1. What amount is 30% of 140? Substitute the values into the percent equation, amount = % × base . Enter 30% as a decimal. a=☐ ×140 Multiply to find a. Answer:	
3. 34 is 50% of what number? Substitute the values into the percent equation, amount = % × base . Enter 50% as a decimal. 34 = ☐ × b Solve the equation for b by dividing each side of the equation by the decimal in front of b. Answer:	

Example	Notes
5. 45 is what percent of 180? Substitute the values into the percent equation, amount = %×base , letting p stand for the unknown percent. $\square = p \times \square$ Solve for p. Answer:	
6. What percent of 30 is 10? Round to the nearest tenth of a percent. Answer:	

Helpful Hints
Usually the base appears after the word "of."

In math "is" translates to an equal sign, and "of" translates into multiplication.

Concept Check
1. State the percent equation and how it can be used to find each of one of the three parts, provided the other two parts are given.

Practice
Use the percent equation to find the following. Round to the nearest tenth of a percent as needed.

2. What amount is 40% of 150?

3. 24 is 20% of what number?

4. 32 is what percent of 160?

5. 25 is what percent of 86?

Ratios, Rates, and Percents
Topic 10.8 The Percent Proportion

Vocabulary
amount   •   percent proportion   •   ratio   •   fraction

1. The _____ states "the ratio of the amount to the base is equal to the ratio of the percent $\frac{p}{100}$."

Step-by-Step Video Notes
Watch the Step-by-Step Video lesson and complete the examples below.

Example	Notes
1. What amount is 20% of 85? Substitute the values into the percent proportion, $\dfrac{amount}{base} = \dfrac{p}{100}$. $\quad \dfrac{a}{\Box} = \dfrac{20}{100}$ Cross multiply. $100(a) = \boxed{}$ Solve for a by dividing each side of the equation by 100. Answer:	
2. 21 is 10.5% of what number? Substitute the values into the percent proportion, $\dfrac{amount}{base} = \dfrac{p}{100}$. $\quad \dfrac{\Box}{b} = \dfrac{\Box}{100}$ Cross multiply. $\left(\boxed{}\right)b = \left(\boxed{}\right)(100)$ Solve for b. Answer:	

Example	Notes
3. 83 is what percent of 332? Substitute the values into the percent proportion, $\dfrac{amount}{base} = \dfrac{p}{100}$. Solve for p. Answer:	
4. 4 is what percent of 48? Round to the nearest tenth of a percent. Answer:	

Helpful Hints

The percent proportion is an alternative to the percent equation.

Usually the base appears after the word "of."

Concept Check

1. State the percent proportion and how it can be used to find each of one of the three parts, provided the other two parts are given.

Practice

Use the percent proportion to find the following.

2. What amount is 30% of 160?

3. 33 is 11% of what number?

Round to the nearest tenth as necessary.

4. 96 is what percent of 384?

5. 6 is what percent of 96?

Ratios, Rates, and Percents
Topic 10.9 Percent Applications

Vocabulary

amount   •   percent   •   base   •   fraction

1. The percent equation states "the amount is equal to the percent times the
    _____."

2. The percent proportion states "the ratio of the _____ to the base is equal to the
    ratio of the percent $\dfrac{p}{100}$."

Step-by-Step Video Notes
Watch the Step-by-Step Video lesson and complete the examples below.

Example	Notes
1. Eli purchased a wrist watch for \$60. If the sales tax rate is 7%, how much sales tax did Eli pay? What was the total cost of the watch? Substitute the values into the application percent equation: $\begin{pmatrix}\text{amount of} \\ \text{sales tax}\end{pmatrix} = \begin{pmatrix}\text{tax rate} \\ \text{as a decimal}\end{pmatrix} \times \begin{pmatrix}\text{cost of} \\ \text{item}\end{pmatrix}$ $a = (.07) \times \square$ Multiply to solve for a. $a = \boxed{}$ Substitute values and perform addition to find total cost. $\begin{pmatrix}\text{total cost} \\ \text{of item}\end{pmatrix} = \begin{pmatrix}\text{amount of} \\ \text{sales tax}\end{pmatrix} + \begin{pmatrix}\text{cost of} \\ \text{item}\end{pmatrix}$ Answers:	

Example	Notes
4. Calculate the amount of interest and the total amount paid for a loan of $2,000 for 3 years at 4% annual rate interest. First we calculate the amount of interest: Amount of Interest = Principal × Rate × Time Next we calculate the total amount paid: Total Amount = Principal + Amount of Interest Answer:	

Helpful Hints

If you know the percent and the base when solving a percent application problem, you should use the percent equation.

Remember to use the percent as a decimal in the percent equation.

Concept Check

1. How does finding the amount of sales tax of an item make use of the percent equation?

Practice

Samantha wants to buy a $40 sweater. Find the following.

2. What is the amount of sales tax if the sales tax rate is 5%?

4. What is the amount of discount if there is a 15% discount sale?

3. What is the total cost of the sweater if the sales tax rate is 5%

5. What is the sale price if Samantha buys the sweater during the 15% discount sale?

Name: _____   Date: _____

Instructor: _____   Section: _____

Introduction to Geometry
Topic 11.1 Lines and Angles

Vocabulary
point • line • line segment • ray • angle • measurement • right angle
acute angle • straight angle • obtuse angle • complementary • supplementary

1. A(n) _____ is a portion of a line with two endpoints.

2. A(n) _____ is a portion of a line with one endpoint.

3. A(n) _____ is an angle that measures 90°.

4. Two angles are _____ if the sum of their measures is 180°.

Step-by-Step Video Notes
Watch the Step-by-Step Video lesson and complete the examples below.

Example	Notes
1. Identify the following figure as a line, line segment, or a ray and give the name. ● A ————● B ———▶ Use the endpoint(s) and the appropriate symbol to name the figure. Answer:	
4b. Identify the angle shown below as a straight angle, right angle or neither. **180°** ◀——● A ——● B ——● C ——▶ Answer:	

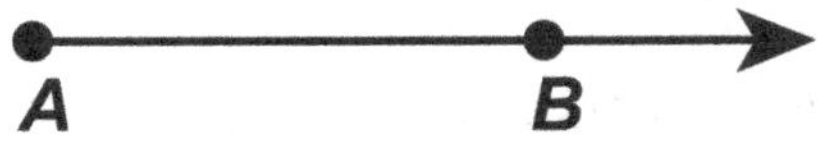

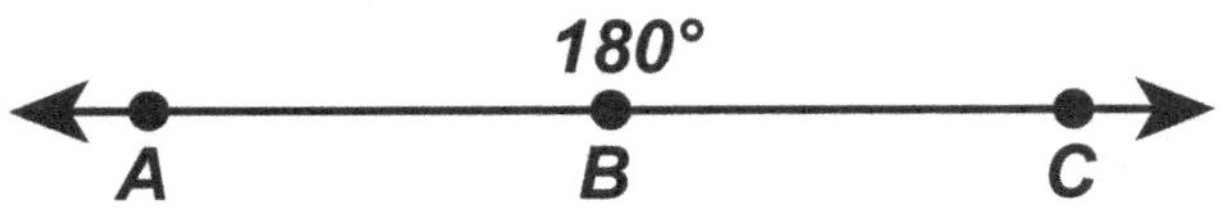

Example	Notes
5a. Identify the given angle as acute, obtuse, or neither. Review the definitions of acute angle and obtuse angle. Answer:	
7a. Find the complement of a 65° angle. Two angles are complementary if the sum of their angles is $\boxed{\,^\circ}$. Subtract 65° from the sum. Answer:	

Helpful Hints

If two complementary angles are adjacent, they will form a right angle.

If two supplementary angles are adjacent, they will form a straight angle.

Concept Check

1. Describe how 90° and 180° are used to define the following angles: acute, right, obtuse, straight, complementary and supplementary.

Practice

Identify the type of angle described as acute, right, or obtuse.

2. An angle measuring 65°

3. An angle measuring 90°

Find the following angles.

4. The supplement of 120°

5. The complement of 35°

Name: _____   Date: _____

Instructor: _____   Section: _____

Introduction to Geometry
Topic 11.2 Figures

Vocabulary

triangle	• right triangle	• acute triangle	• obtuse triangle
polygon	• quadrilateral	• parallel lines	• rectangle
square	• trapezoid	• parallelogram	• parallel lines

1. A(n) _____ is a four-sided geometric figure.

2. A(n) _____ is a quadrilateral with opposite sides that are equal in length and four angles that are right angles.

3. A(n) _____ is a quadrilateral with only one pair of opposite sides that are parallel.

Step-by-Step Video Notes
Watch the Step-by-Step Video lesson and complete the examples below.

Example	Notes
1. If two angles of a triangle measure 55° and 70°, find the measure of the third angle. Then identify the triangle as acute, right, or obtuse. Find the sum of the two given angles. ☐° Subtract this sum from 180. $180 - \boxed{} = \boxed{}$ Is there an obtuse angle? _____ Is there a right angle? _____ Are all the angles acute? _____ Answer:	

Example	Notes
2. Identify the figure below. Answer:	
3. Identify the figure below. Answer:	

Helpful Hints
The sum of the angles of any triangle is 180°.

Every square is also a rectangle, but every rectangle is not a square.

Concept Check
1. What are the similarities and differences among these quadrilaterals: rectangle, square, trapezoid, parallelogram, diamond and kite.

Practice

2. Find the third angle of a triangle having angles of 100° and 50°.

4. Identify a triangle with angles of 95°, 45°, and 40° as acute, right or obtuse.

3. Identify the figure below.

5. Identify the figure below.

Introduction to Geometry
Topic 11.3 Perimeter - Definitions and Units

Vocabulary
polygon   •   distance   •   perimeter   •   parallelogram

1. The _____ of a figure can be found by adding the lengths of all its sides.

Step-by-Step Video Notes
Watch the Step-by-Step Video lesson and complete the examples below.

Example	Notes
1. Find the perimeter of the figure. 6 inches 8 inches 4 inches Add the lengths of the three sides of the figure. 6 inches + ☐ inches + ☐ inches = ☐ inches Answer:	
2. Find the perimeter of the figure. 3 cm 3 cm 3 cm 3 cm 3 cm Add the lengths of the five sides of the figure. Answer:	

Example	**Notes**
3. Find the perimeter of the figure. 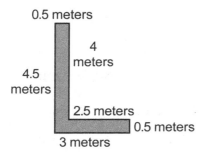 Answer:	

Helpful Hints
Perimeter is always measured in units of length.

Concept Check
1. What is the similarity in finding the perimeter of a rectangle and a triangle?

Practice
2. Which of these units, if any, could represent perimeter? Square feet, miles, or cubic meters

4. Find the perimeter of the figure.

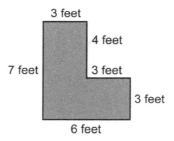

3. Find the perimeter of the figure.

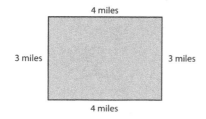

5. Find the perimeter of the figure.

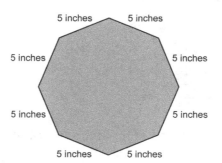

Introduction to Geometry
Topic 11.4 Finding Perimeter

Vocabulary
angle   •   square   •   perimeter   •   rectangle

1. The _____ of a figure can be found by adding the lengths of all its sides.

Step-by-Step Video Notes
Watch the Step-by-Step Video lesson and complete the examples below.

Example	Notes
1. Find the perimeter of the triangle below.	

3.5 cm   2.5 cm

4 cm

Use the perimeter formula: $P = a + b + c$, where a, b, and c represent the sides of the triangle.

$P = 3.5 \text{ cm} + \boxed{} \text{ cm} + \boxed{} \text{ cm}$

Find the sum.

Answer:

2. Find the perimeter of the rectangle below.

30 feet

40 feet

Use the perimeter formula: $P = 2l + 2w$, where l represents length and w represents width.

Answer:

Example	Notes
3. Find the perimeter of the figure. 14 m 14 m Use the formula: $P = 4s$, where s represents the side length of the square. Answer:	
4. Find the perimeter of a triangular garden with side lengths of 5 feet, 7 feet, and 10 feet. Answer:	

Helpful Hints

Perimeter is always measured in units of length.

The perimeter can be found by adding the lengths of the sides of a figure, but formulas for a triangle, a square and a rectangle can be used.

Concept Check

1. What are the perimeter formulas for a triangle, a rectangle, and a square?

Practice

Use the appropriate perimeter formula to find the perimeter of the figures shown or described below.

2.

6 cm   6 cm

10 cm

4.
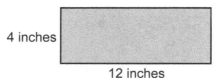

4 inches

12 inches

3. A rectangular desk with length 4 feet and width 2.5 feet.

5. A square bandana with side lengths of 8 inches.

Introduction to Geometry
Topic 11.5 Area - Definitions and Units

Vocabulary

square unit   •   area   •   perimeter   •   square inches

1. The _____ of a figure is the measure of the surface inside the figure, which is measured in square units.

Step-by-Step Video Notes
Watch the Step-by-Step Video lesson and complete the examples below.

Example	**Notes**
1. How many square units are needed to cover the figure below completely? 4 inches 3 inches 3 inches 4 inches The small squares measuring 1 in.×1 in. are square units. Count the number of square units in this figure. Answer:	
3. Draw two different rectangles each of which has an area of 6 square units.	

Example	Notes

5. Find the area of a right triangle with a base and height of 4 in.

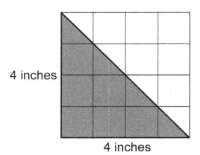

4 inches

4 inches

Add the half squares to make whole squares.

Answer:

Helpful Hints

Area is always measured in square units of length.

Two figures can have different shapes, but have the same area.

Concept Check

1. How is a grid of square units used to determine the area of a figure?

Practice

2. Find the area of a right triangle with a base and height of 3 in.

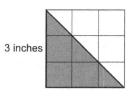

3 inches

3 inches

3. How many square units are needed to cover the figure below completely?

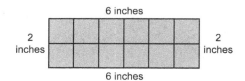

6 inches

2 inches   2 inches

6 inches

4. Draw a rectangle which has an area of 12 square units.

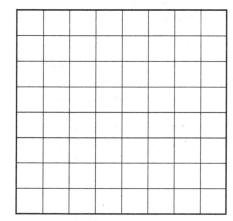

Introduction to Geometry
Topic 11.6 Finding Area

Vocabulary
triangle • area • perimeter • quadrilateral • parallelogram • rectangle • square

1. _____ is the measure of the surface inside a figure.

Step-by-Step Video Notes
Watch the Step-by-Step Video lesson and complete the examples below.

Example	Notes
1. Use the formula for area of a rectangle to find the area of the following rectangle. 9 inches 6 inches Substitute 9 for *l* and 6 for *w* in the formula. $A = lw = \left(\square \text{ in.}\right)\left(\square \text{ in.}\right) = \square \text{ in.}^2$ Answer:	
2. Use the formula for area of a square to find the area of a square deck with side lengths of 3 meters. 3 m 3 m Substitute 3 for *s* in the formula. $A = s^2$ Answer:	

Example	Notes
3. Use the formula for area of a triangle to find the area of a right triangle with a base of 4 cm and a height of 3 cm. 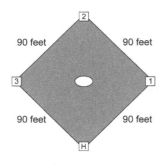 Answer:	
5. Find the area of the infield of a major league baseball diamond, which is a square whose sides are 90 feet long.	

Helpful Hints
Remember to include the square units of length when finding area of a geometric figure.

Concept Check
1. What are the formulas for finding the area of a rectangle, a square, a parallelogram, and a triangle?

Practice
Find the area of the following.
2. A rectangle whose length is 12 inches and width is 3 inches

4. A parallelogram whose base is 23 feet and height is 8 feet

3. A right triangle with a base of 12 cm and a height of 5 cm

5. A diamond logo on a shirt which is a square whose sides are 8 cm long.

Introduction to Geometry
Topic 11.7 Understanding Circles

Vocabulary
circle   •   diameter   •   area   •   radius   •   circumference   •   pi

1. A _____ is a figure in which all points on the circle are the same distance from a fixed point called the center.

2. The _____ of a circle is the distance from the center to a point on the circle.

3. The _____ of a circle is the distance around the circle.

Step-by-Step Video Notes
Watch the Step-by-Step Video lesson and complete the examples below.

Example	Notes
1. Use the appropriate formula to find the radius of a pizza with a diameter of 10 inches. Use the formula $r = \frac{1}{2}d$, where d is the diameter. $r = \frac{1}{2}\left(\square \text{ inches}\right) = \square \text{ inches}$ Answer:	
2. A neighbor purchased a 14-foot circular trampoline for his children. Does this situation involve the diameter or the radius? 14 foot Review the definitions for diameter and radius. Answer:	

Example	Notes
3. In softball, the circle around the pitcher's mound is drawn 6 feet from where the pitcher stands. Does this situation involve the diameter or the radius? 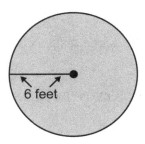 Answer:	

Helpful Hints
The radius is always smaller than the diameter; the radius is half of the diameter.

The radius extends from the center of a circle, while the diameter passes through the center.

Concept Check
1. Define these parts of a circle: radius, diameter and circumference.

Practice
Does the situation described involve the diameter of the radius?

2. A toy helicopter has moveable blades that are 5 inches long.

3. A child's circular board game folds along the center with a measure of 15 inches.

Find the radius in the following situations.

4. A circular garden with a diameter of 4 feet.

5. A circular drum with a diameter of 12 inches.

Introduction to Geometry
Topic 11.8 Finding Circumference

Vocabulary
diameter   •   pi   •   radius   •   circumference   •   area

1. $\pi = \dfrac{C}{d}$, where C is the circumference and d is the _____.

2. The _____ of a circle is the distance around the circle.

Step-by-Step Video Notes
Watch the Step-by-Step Video lesson and complete the examples below.

Example	Notes
1. Find the circumference of a circle with a radius of 8 inches. Find the exact answer in terms of π. Then find the approximation using 3.14 for π. Use the formula $C = 2\pi r$. $C = 2\pi \left(\Box \text{ in.} \right) = \Box\, \pi \text{ in.}$ Use 3.14 for π. $C = \Box (3.14) \text{ in.} = \Box \text{ in.}$ Answer:	
2. Find the circumference of a circle with a diameter of 5 cm. Find the exact answer in terms of π. Then find the approximation using 3.14 for π. $C = \pi \left(\Box \text{ cm} \right) = \Box\, \pi \text{ cm}$ Use 3.14 for π. Answer:	

Example	Notes
5. The earth's equator forms a circle. Estimate the number of miles a ship would have to travel if it went around the earth at the equator. Use 7900 miles as an approximation of the earth's diameter. Use 3.14 for π. Answer:	

Helpful Hints

Exact answers for circumference are left in terms of π.

Approximations for circumference use $\dfrac{22}{7}$ or 3.14 for the value of π.

Concept Check

1. What are the two formulas for finding the circumference of a circle?

Practice

Find the circumference of the following circles. Find the exact answer in terms of π. Then find the approximation using 3.14 for π.

2. A circle with a radius of 6 meters

3. A circle with a diameter of 7 inches

4. The bottom of an empty farm silo forms a circle. The diameter of the silo is 18 feet.

5. The lid to a trash barrel is a circle of radius 1.5 feet.

Introduction to Geometry
Topic 11.9 Finding Area - Circles

Vocabulary
diameter   •   area   •   radius   •   circumference

1. The _____ of a figure is the measure of the surface inside the figure.

Step-by-Step Video Notes
Watch the Step-by-Step Video lesson and complete the examples below.

Example	Notes
1. Find the area of a circle with a radius of 6 feet. Use the approximate value of 3.14 for π. 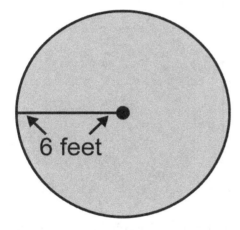 Use the formula $A = \pi r^2$, where A is the area and r is the radius. Enter the radius. $A = (3.14)(\square\ \text{feet})^2$ Square the radius. $A = (3.14)(\square\ \text{square feet})$ Answer:	

Example	Notes
2. Find the area of a circle with a diameter of 6 meters. Use the approximate value of 3.14 for π. 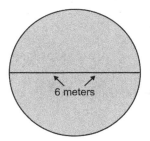 Find the radius using $r = \dfrac{1}{2}d$ Next use the formula $A = \pi r^2$. Answer:	
3. Find the area of a circle with a diameter of 7 feet. Use the approximate value of $\dfrac{22}{7}$ for π. Answer:	

Helpful Hints
Remember that area is measured in square units.

If the diameter is given when finding the area of a circle, it must be divided by 2 to get the radius.

Concept Check
1. What is the formula for finding the area of a circle?

Practice
Find the area of the following circles. Use the approximate value of 3.14 for π.
2. A circle with a radius of 10 feet   4. A circle with a diameter of 3.5 cm

3. A circle with a radius of 2.5 meters   5. A circle with a diameter of 10 mi

Name: _____   Date: _____

Instructor: _____   Section: _____

More Geometry
Topic 12.1 Volume - Definitions and Units

Vocabulary
volume   •   cube   •   box   •   area

1. A _____ is a rectangular box in which every side is a square.

2. The _____ of a three-dimensional figure is the amount of space inside the figure.

Step-by-Step Video Notes
Watch the Step-by-Step Video lesson and complete the examples below.

Example	Notes
1. Find the volume of a box with a length of 4 inches, a width of 3 inches, and a height of 4 inches. How many cubes will it take to fill the bottom layer? $4 \times 3 = \boxed{}$ How many layers will fill the box? $\boxed{}$ Multiply the number of layers by the number of cubes in each layer. Answer in cubic units. Answer:	

Example	Notes
2. Find the volume of a cube that measures 4 cm on each side. 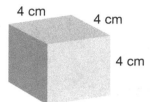 How many 1 cm cubes will it take to fill the bottom layer? ☐ Answer:	
3. Find the volume of a box with length 5 m, width 1 m, and height 3 m. 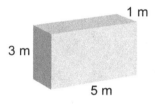 Answer:	

Helpful Hints

Volume involves working with three dimensions; length, width, and height.

Volume is measured in cubic units. You can write out the cubic units for the label (cubic meters), or use the exponent and the abbreviation (m^3).

Concept Check
1. How does measuring area help to find volume for a cube or a box?

Practice

Find the volume of a cube with the given side measure.

2. 6 in.

3. 9 feet

Find the volume of a rectangular box with the given dimensions.

4. length 4 ft, width 2 ft, height 7 ft

5. length of 3 m, width of 6 m, height of 5 m

More Geometry
Topic 12.2 Finding Volume

Vocabulary
cylinder   •   sphere   •   pi   •   formula   •   radius

1. Using a _____ is a faster and more practical way to find the volume of a three-dimensional object than by counting unit cubes.

Step-by-Step Video Notes
Watch the Step-by-Step Video lesson and complete the examples below.

Example	Notes
1. Use the formula to find the volume of a rectangular box with a length of 12 inches, a width of 10 inches, and a height of 5 inches. $V = lwh = 12 \times 10 \times 5 = \boxed{}$ Answer:	
2. Use the formula to find the volume of a cube that measures 2.5 km on each side. Round to the nearest hundredth. $V = s^3 = (\boxed{})^3$ Answer:	

Example	Notes
3. Use $V = \pi r^2 h$ and the approximate value of 3.14 for π to find the volume of a cylinder with radius of 3 inches and a height of 5 inches. Answer:	
4. Use the formula and the approximate value of 3.14 for π to find the volume of a sphere with a radius of 6 mm. Answer:	

Helpful Hints

Use the formulas $V = lwh$, $V = s^3$, $V = \pi r^2 h$, and $V = \dfrac{4\pi r^3}{3}$ to find the volumes of rectangular boxes, cubes, cylinders, and spheres, respectively.

Concept Check
1. How does using a formula make it simpler to find the volume of an object?

Practice
Find the volume of the figure with the given dimensions.
2. a cube with a side length 8 ft

3. a rectangular box with length 8 cm, width 12 cm, height 11 cm

Find the volume of the figure with the given dimensions. Use 3.14 for π.
4. a cylinder with a radius of 8 in. and a height of 9 in.

5. a sphere with a radius of 1 meter

More Geometry
Topic 12.3 Square Roots

Vocabulary
perfect square   •   square root   •   area of a square   •   radical   •   irrational numbers

1. If $a^2 = b$, then a is the principal _____ of b and is denoted $\sqrt{b}$.

2. A _____ is a whole number that can be written as a whole number multiplied by itself.

Step-by-Step Video Notes
Watch the Step-by-Step Video lesson and complete the examples below.

Example	Notes
1–4. Determine whether each number is a perfect square. 81 50 64 27	
5. Find the square root of 36. Finding a square root is the reverse operation of squaring a number. The square root of 36 is ☐ , because ☐² = 36. Answer:	
6. Simplify $\sqrt{144}$. Answer:	

Example	Notes
8. The area of a square is 49 cm^2. What is the length of each side? The area of a square is found using the formula $A = s^2$, where s is the length of a side. Answer:	
10. Approximate $\sqrt{75}$ by finding the two consecutive whole numbers the square root lies between. Answer:	

Helpful Hints

If a whole number is not a perfect square, use a calculator, or approximate the square root by finding two perfect squares that whole number lies between.

Irrational numbers are non-terminating, non-repeating decimals.

Concept Check

1. Can you find the perimeter of a square if you know its area? Explain.

Practice

2. Find the square root of 100.

3. Simplify $\sqrt{225}$.

4. Use a calculator to approximate $\sqrt{154}$. Round to the nearest hundredth.

5. Approximate $\sqrt{130}$ by finding the two consecutive whole numbers the square root lies between.

More Geometry
Topic 12.4 The Pythagorean Theorem

Vocabulary

Pythagorean Theorem • hypotenuse • leg • diagonal

1. The longest side of a right triangle, opposite the right angle, is the _____.

2. The _____ states that in a right triangle, the sum of the squares of the legs is equal to the square of the hypotenuse, or $\text{leg}^2 + \text{leg}^2 = \text{hypotenuse}^2$.

Step-by-Step Video Notes
Watch the Step-by-Step Video lesson and complete the examples below.

Example	Notes
1. Find the length of the hypotenuse of a right triangle with legs that measure 3 cm and 4 cm. To find the hypotenuse, use the formula $\text{hypotenuse} = \sqrt{\text{leg}^2 + \text{leg}^2}$. $\text{hypotenuse} = \sqrt{3^2 + 4^2} = \sqrt{9 + \square} = \sqrt{\square} = \square$ Answer:	
2. One leg of a right triangle measures 9 inches, and the hypotenuse measures 15 inches. Find the length of the other leg. To find the length of the missing leg, use the formula $\text{leg} = \sqrt{\text{hypotenuse}^2 - \text{leg}^2}$. Substitute for the hypotenuse and leg. $\text{leg} = \sqrt{15^2 - \square^2} = \sqrt{\square}$ Answer:	

Example	Notes
3. Find the missing side of the right triangle. Round to the nearest tenth. 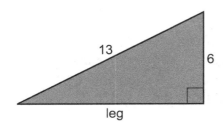 Identify the formula to use. Substitute the given values into the formula. Find the missing side. Answer:	
4. A standard computer monitor has a length of 20 inches and a width of 15 inches. What is the measure of the diagonal of the monitor? The monitor is a rectangle. The diagonal divides the rectangle into 2 identical right triangles. The length and width are the legs, and the diagonal is the hypotenuse. Identify the formula to use. Answer:	

Helpful Hints
The Pythagorean Theorem works with right triangles or the many real-life situations where a right triangle can be applied to help find a distance. Be sure to identify whether the side of the triangle you are trying to find is a leg or the hypotenuse, and use the appropriate formula. Quite often, the Pythagorean Theorem is stated as $a^2 + b^2 = c^2$.

Concept Check
1. How do you find the length of the diagonal of a rectangle?

Practice
Find the hypotenuse of a right triangle with the given length of the legs.
2. Leg = 5 ft Leg = 12 ft

3. Leg = 16 m Leg = 12 m

Find the length of the missing leg of a right triangle given the hypotenuse and a leg.
4. Leg = 24 in. Hypotenuse = 25 in.

5. Leg = 24 cm Hypotenuse = 26 cm

Name: _____   Date: _____

Instructor: _____   Section: _____

More Geometry
Topic 12.5 Similar Figures

Vocabulary
similar figures   •   proportion   •   corresponding sides   •   ratio

1. _____ have the same shape and corresponding angles with the same measure.

Step-by-Step Video Notes
Watch the Step-by-Step Video lesson and complete the examples below.

Example	Notes
1. Identify the corresponding sides and set up proportions to determine if the figures are similar. 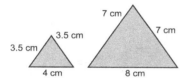 The left and right sides of the smaller triangle measure 3.5 cm. The corresponding sides of the larger triangle measure 7 cm. The ratio is $\dfrac{3.5}{7}$. Compare the bottom side of the smaller triangle to the bottom side of the larger triangle. Is $\dfrac{3.5}{7} = \dfrac{\square}{\square}$?	
2. Identify the corresponding sides and set up proportions to determine if the figures are similar. 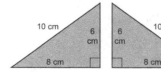 Both figures are right triangles. $\dfrac{6}{6} = \dfrac{\square}{\square} = \dfrac{\square}{\square}$, so the figures _____ similar.	

Example	Notes

4 & 5. Determine if the figures in each pair are similar.

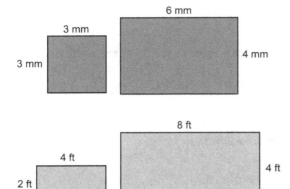

In the first set of rectangles, is $\dfrac{3}{4} = \dfrac{3}{6}$?

In the second set of rectangles, is $\dfrac{2}{4} = \dfrac{\square}{8}$?

Answer:

Helpful Hints
The concept of similar figures applies to many everyday applications such as photography, art, scale models, and maps. If figures are similar, they are proportional, or "to scale."

If the corresponding sides of two triangles are proportional, the triangles are similar. Likewise, two rectangles are similar if their corresponding sides are proportional.

Concept Check
1. Can you think of geometric shapes other than squares that are always similar?

Practice
Determine if the figures are similar.
2. Triangle A with sides 4 m, 5 m, 6m
    Triangle B with sides 8 m, 12 m, 10 m

3. Triangle C with sides 9 ft, 9 ft, 7 ft
    Equilateral Triangle D with sides 9 ft

Can the following pairs of figures be similar?
4. A rectangle and a trapezoid

5. A very small square measured in mm
    and a very large square measured in km

More Geometry
Topic 12.6 Finding Missing Lengths

Vocabulary
corresponding sides • proportion • proportional • unknown

1. Corresponding sides of similar figures are _____.

Step-by-Step Video Notes
Watch the Step-by-Step Video lesson and complete the examples below.

Example	Notes
1 & 2. Write a proportion for the corresponding sides of the similar figures.	

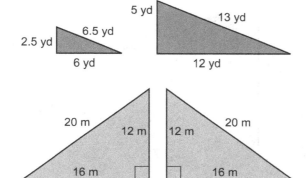

The proportion for the corresponding sides of

the first pair of triangles is $\dfrac{2.5}{5} = \dfrac{\square}{12} = \dfrac{6.5}{\square}$.

Each ratio in the proportion is equal to $\dfrac{1}{\square}$.

In the second pair, the figures have exactly the same shape, therefore they are _____.

Answer:

Example	Notes
3. Find the unknown side length in the similar figures. 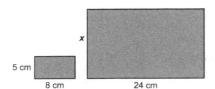 The proportion is $\dfrac{\square}{x} = \dfrac{8}{\square}$. Answer:	

4. Find the unknown side length in the similar figures.

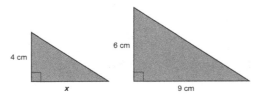

The proportion is $\dfrac{4}{\square} = \dfrac{x}{\square}$.

Answer:

Helpful Hints
If figures are similar, the ratios of the corresponding sides will be equal. Line up the corresponding sides of similar figures in a proportion to find unknown lengths.

Concept Check
1. Is a 30 in. wide by 36 in. tall poster similar to a 4 in. by 6 in. photo?

Practice
Solve for x.
2. A 6 in. by 9 in. rectangle is similar to a 24 in. by x in. rectangle

Find the unknown length.
4. A 5 cm by 12 cm rectangle is similar to a 15 mm by ☐ mm rectangle.

3. A triangle with sides 6 m, 10 m, 8 m is similar to a triangle with sides 9 m, 15 m, and x m.

5. $\dfrac{8 \text{ ft}}{12 \text{ ft}} = \dfrac{10 \text{ ft}}{\boxed{}}$

More Geometry
Topic 12.7 Congruent Triangles

Vocabulary
Side-Side-Side (SSS) • Side-Angle-Side (SAS) • Angle-Side-Angle (ASA) • congruence

1. Two triangles are congruent by _____ if two sides of one triangle are the same length as two sides of the second triangle, and the measure of the angle formed by these two sides in each triangle is equal.

2. Two triangles are congruent by _____ if two angles of one triangle are equal in measure to two angles of the second triangle, and the side between these two angles in each triangle is the same length.

Step-by-Step Video Notes
Watch the Step-by-Step Video lesson and complete the examples below.

Example	Notes
1. Explain why ΔABC and ΔEFD are congruent.	

Answer:

| 2. Explain why ΔSTU and ΔCAB are congruent. | |

Answer:

171

Example	Notes

4. Determine whether the pair of triangles is congruent.

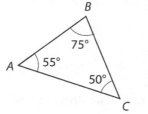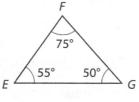

Answer:

Helpful Hints

There are combinations of congruent sides and angles that do not show that triangles are necessarily congruent.

Concept Check

1. Why are two equilateral triangles not necessarily congruent?

Practice

2. Explain why $\triangle ABC$ and $\triangle DEF$ are congruent.

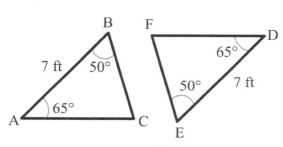

4. Determine whether the pair of triangles is congruent.

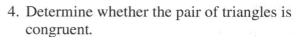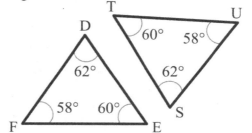

3. Explain why $\triangle STU$ and $\triangle CAB$ are congruent.

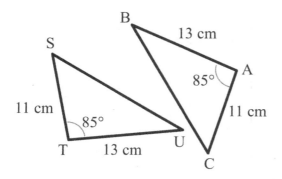

5. Determine whether the pair of triangles is congruent.

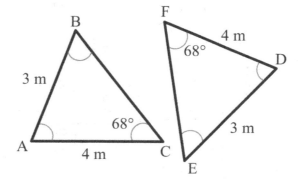

More Geometry
Topic 12.8 Applications of Equations and Geometric Figures

Vocabulary
diameter • circumference • volume • π • Pythagorean Theorem

1. The _____ of a circle is the distance around the circle.

2. The _____ of a circle is the distance across the circle, passing through the center.

Step-by-Step Video Notes
Watch the Step-by-Step Video lesson and complete the examples below.

Example	Notes
1. Find the unknown width of the rectangular solid. $V = 300$ cm³ 6 cm w 10 cm Answer:	
2. Find the length of the guy wire supporting the telephone pole. Wire 15 ft 8 ft Answer:	

Example	Notes
3. A manhole cover has a diameter of 3 ft. A strip of brass runs around the circumference of the cover. What is the length of this brass strip? Round to the nearest hundredth. (Use 3.14 for π). Answer:	
4. A flagpole casts a shadow. At the same time, a small tree casts a shadow. Use the sketch to find the height n of the flagpole. 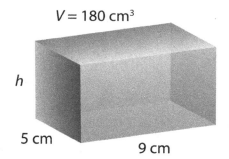 Answer:	

Helpful Hints

When solving problems, first, be sure you understand the problem. Then, create a plan for solving the problem. Next, find the answer by performing all the necessary calculations. Last, check the answer.

Concept Check

1. Tori needed to find the volume of a cylinder with a diameter of 12 inches and a height of 18 inches. She calculated $V = \pi (12)^2 \cdot 18 \approx 8143$ in^3. Find and correct her error.

Practice

2. Find the unknown height of the rectangular solid.

$V = 180$ cm³

h

5 cm

9 cm

3. A tree casts a shadow on the ground that is 48 feet long. At the same time and location, Sharon casts a shadow that is 12 feet long. If Sharon is 5 feet tall, how tall is the tree?

4. A wheel with a diameter of 20 inches rolls twelve times across the ground. How far did the wheel travel? (Use 3.14 for π).

Name: _____   Date: _____

Instructor: _____   Section: _____

Statistics
Topic 13.1 Bar Graphs

Vocabulary
bar graph • double bar graph • measurement • frequency distribution table • histogram

1. A _____ is a special type of bar graph where the width of the bar represents an interval, or range of numbers.

Step-by-Step Video Notes
Watch the Step-by-Step Video lesson and complete the examples below.

Example	Notes
2.1 Use the bar graph below to answer the question. How many Country Music Association Awards did Loretta Lynn win?	

Loretta Lynn
Country Music Association Awards

Type of Awards

The height of each bar represents the number in each category. Enter these numbers from the graph.

Number of awards = $4 + \square + \square$

Answer:

2.2 Use the bar graph in example #2.1 to answer the question. In which category did she win the fewest awards?

Which bar is the lowest? _____

Example	Notes

5. In an Intro to Music class, letter grades are distributed based on the following scale: score of 90-100 an A, 80-89 a B, 70-79 a C, 60-69 a D, and below 60 is an F. Enter the tally and frequency in the table below for a class with scores of 78, 69, 82, 95, 92, 80, 47, 89, 81, and 99. Then use the table to create a histogram displaying the same data.

Intro To Music Grade Distribution		
Grade Intervals	Tally	Frequency
90 – 100		
80 – 89		
70 – 79		
60 – 69		
Below 60		

Helpful Hints

A bar graph can be used to display data over time, to compare amounts, or show how often a particular amount will occur.

The height of the rectangular bar indicates the number in each category in a bar graph.

Concept Check

1. Name two similarities and one difference between a bar graph and a histogram.

Practice

Use the bar graph below to answer the following.

2. How many softball team members voted?

Softball Team Captain Vote

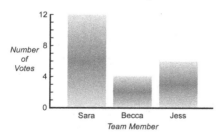

3. Which player received the fewest votes?

Fill in the frequency distribution table for the given data below.

4. The Human Resources Manager at a company tracks employee sick days in the categories of 0-2, 3-5, 6-8, 9-11, and 12 or more. The art department employees used sick days of 4, 6, 2, 3, 9, 0, 1, 0, 3, 2, 4, 7, 0, 1, and 15.

Employee Sick Days Distribution		
# of days used	Tally	Frequency
0 – 2		
3 – 5		
6 – 8		
9 – 11		
12 or more		

Statistics
Topic 13.2 Line Graphs

Vocabulary
double line graphs   •   line graphs   •   data points   •   histograms

1. _____ can display data, show trends or patterns in data over time, show how often a particular data value occurs, compare two or more types of data, and use data points that are connected with straight line segments.

Step-by-Step Video Notes
Watch the Step-by-Step Video lesson and complete the examples below.

Example	Notes
1.1 Use the line graph below to answer the question. Has the average life expectancy of humans increased or decreased since 1920? **Average Life Expectancy of Humans from 1920 to 2000** *Years of Life* — 80, 70, 60 *Decades in the 1900s* — 1920 1940 1960 1980 2000 Answer:	
1.2 Use the line graph in example #1.1 to answer the question. During which 20-year interval did the life expectancy increase the most? Answer:	

Example	Notes
3.1 Use the double line graph below to answer the question. Which countries had the same life expectancy in for 2000 and 1998? 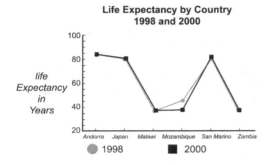 Answer:	

Helpful Hints

A line graph is usually used to show trends or patterns of data over time, while a double line graph is used to compare more than one set of data on a graph.

Concept Check

1. State at least two differences between a bar graph and a line graph.

Practice

Use the line graph below to answer the following.

2. Have the vehicle accidents in Springfield increased or decreased since 1960?

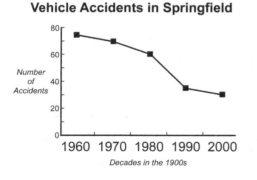

3. During which decade was the decrease in accidents the most?

Use the double line graph below to answer the following.

4. During which decade were there more vehicle accidents in Newton than in Springfield?

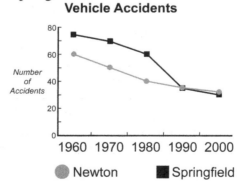

Statistics
Topic 13.3 Circle Graphs

Vocabulary

line graph   •   circle graph   •   amounts   •   percents

1. A _____, also called a pie chart, is drawn to show actual amounts or percents and is used to show how one set of data is divided.

Step-by-Step Video Notes
Watch the Step-by-Step Video lesson and complete the examples below.

Example	Notes
1.1 Use the circle graph below to answer the question. What does Nikeshia plan to spend the most money on? **Average Costs of School** Books — $400 Tuition — $1200 Housing — $800 Food — $800 Other — $400 Answer:	
1.2 Use the circle graph in example #1.1 to answer the question. What total amount does Nikeshia plan to spend on food and housing? Find the sum of the amount from the food category and the housing category. Answer:	

Example	Notes
3.1 Use the circle graph below to answer the question. Which assignment type counts the most toward the student's final grade? 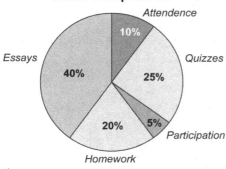 **Breakdown of Writing Class Grade Components** Answer:	

Helpful Hints
Circle graphs are most often used to display relationships that compare parts to a whole.

In circle graphs the sum of all of the parts must equal the total amount or 100%.

Concept Check
1. What is different about a circle graph when compared to a bar graph or line graph?

Practice
Use the circle graphs to answer the following questions.

2. What eye color is the least frequent in this class of students?

4. What percentage of students in this class has blue eyes?

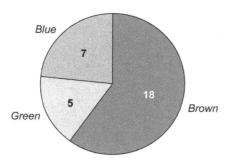

Student Eye Color

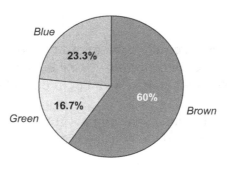

Student Eye Color

3. How many students are in this class?

Statistics
Topic 13.4 Mean

Vocabulary
measures of central tendency   •   statistics   •   mean   •   data

1. _____ is information about a group or topic, often consisting of a set of numbers.

2. The _____ is the sum of the values in the list of numbers divided by the number of values.

Step-by-Step Video Notes
Watch the Step-by-Step Video lesson and complete the examples below.

Example	Notes
1. Calculate the mean for the following list of numbers. 3, 5, 7, 2, 11, 6, 8, 2, 10 Find the sum of all of the values in the list. $3 + 5 + \square + 2 + 11 + 6 + 8 + \square + 10 = \square$ There are 9 values in the list. Divide the sum by the number of values in the list. $\square \div 9 = \square$ Answer:	
2. Calculate the mean of the following list of test scores. 93, 86, 95, 98, 82 Find the sum. $\square$ There are $\square$ values in the list. Divide the sum by the number of values in the list. Answer:	

Example	Notes
3. Eli saved a portion of his allowance for 6 weeks. He saves $14 the first week, $12 the second week, $6 the third week, and $4, $10, and $2 for the remaining weeks. On average, how much did he save each week? Find the average, which is the same as the mean, saved over the six weeks.	

Helpful Hints

Mean is a measure of central tendency because it represents an entire list of numbers as well as the "center" of the data.

The mean is often referred to as the average.

Concept Check

1. Describe how to find the mean of a list of numbers.

Practice

Find the mean of each list of numbers.

2. 10, 8, 4, 7, 5, 8, 10, 4

4. The five players on Jamison's basketball team collected donations for new team uniforms. Jamison collected $18. His teammates collected $12, $34, $26 and $10. On average, how much did each player collect?

3. 55, 42, 33, 61, 30

Statistics
Topic 13.5 Median

Vocabulary

median • statistics • mean • data

1. _____ is the study and use of data.

2. The _____ of a set of ordered numbers is the middle number.

Step-by-Step Video Notes
Watch the Step-by-Step Video lesson and complete the examples below.

Example	Notes
1. Find the median of the following list of numbers. 3, 5, 7, 2, 11, 6, 8, 2, 10 Write the numbers in order from smallest to largest. 2, ☐, 3, 5, ☐, ☐, 8, 10, ☐ Does the list contain an odd number or even number of values? Odd; there are 9 values. Mark off one number from the beginning of the list and from the end of the list. 2̸, 2, 3, 5, ☐, ☐, 8, 10, 1̸1̸ Repeat this step. 2̸, 2̸, 3, 5, ☐, ☐, 8, 1̸0̸, 1̸1̸ Repeat this step until the middle number is reached. Answer:	

Example	Notes
2. Find the median of the following list of numbers. 44, 32, 31, 56, 77, 65 Order the numbers. Since there are an even number of values, the median is the average of the two middle numbers. Find this average. $44 + \boxed{} = \boxed{}$ $\boxed{} \div 2 = \boxed{}$ Answer:	
3. A poll of nine students shows the numbers of Facebook friends they have are 395, 486, 166, 430, 172, 159, 723, 582, and 319. Find the median number of Facebook friends for these nine students. Answer:	

Helpful Hints
Remember that the values must be ordered before finding the median of a list of numbers.

Concept Check
1. How do you find the median of a list that has an even number of values? How does this differ from the process of finding the median of list that has an odd number of values?

Practice
Find the median of each list of numbers.

2. 10, 8, 4, 7, 5, 8, 10

4. Sarah's math test scores for the first term were 95, 92, 81, 98, 84, 86, and 94. Find the median of Sarah's math test scores.

3. 55, 43, 33, 61, 30, 41

Statistics
Topic 13.6 Mode

Vocabulary

median   •   statistics   •   data   •   mode

1. The _____ of a list of numbers is the number that occurs most often in the list.

Step-by-Step Video Notes
Watch the Step-by-Step Video lesson and complete the examples below.

Example	Notes
1. Find the mode(s) of the following list of numbers. 3, 5, 7, 2, 11, 6, 8, 2, 10 Write the numbers in order from smallest to largest. 2, 2, 3, ☐, 6, ☐, ☐, 10, 11 Choose the number(s) that occur the most often. Answer:	
2. Find the mode(s) of the following list of numbers. 44, 32, 31, 56, 77, 65, 65, 44, 65 Order the numbers. 31, ☐, 44, ☐, ☐, ☐, 65, 77 Answer:	
3. Find the mode(s) of the following list of numbers. 325, 612, 367, 692, 451 Note that a list of numbers can have more than one mode or no mode. Answer:	

Example	Notes
5. According to the National Weather Service, the high temperatures (in °F) for the first two weeks of June 2010 were recorded as 105, 106, 109, 112, 116, 119, 119, 115, 111, 107, 106, 103, 104, and 110. Find the mode(s) of the high temperatures for this period. Answer:	

Helpful Hints

It is not a required step to order the numbers when finding the mode, but it can be helpful.

The mode is not just any number that occurs more than once, it is the number that occurs more often than the rest.

Remember that there can be one mode, more than one mode, or no mode.

Concept Check

1. How is the mode different than the other two measures of central tendency, the mean and the median?

Practice

Find the mode(s) of each list of numbers.

2. 65, 35, 40, 55, 40, 30, 70

4. The students in a kindergarten class each have a box for storing crayons and pencils. The number of crayons in the boxes is 8, 6, 7, 6, 2, 5, 7, 4, 3, and 8. Find the mode(s) of the crayons in the boxes.

3. 10, 8, 4, 7, 5, 8, 10

Statistics
Topic 13.7 Introduction to Probability

Vocabulary

probability   •   favorable outcomes   •   possible outcomes

1. _____ measure(s) the likelihood that a given event will occur.

Step-by-Step Video Notes
Watch the Step-by-Step Video lesson and complete the examples below.

Example	Notes
1. Draw a tree diagram showing the possible outcomes of rolling a single die. Answer:	
3. Find the probability of rolling a 4 on a single die. Answer:	
7–9. A bag contains 6 red marbles, 7 blue marbles, 2 black marbles, and 5 green marbles. Find the probability of picking a blue or black marble. Find the probability of picking a marble of any color. Find the probability of picking a yellow marble.	

Example	Notes
12. A board game uses a colored spinner to determine where a player will move on his or her next turn. 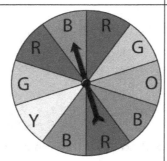 Find the probability of the spinner landing on a primary color (red, yellow, or blue). Answer:	

Helpful Hints

Probability can be defined as $\text{Probability} = \dfrac{\text{Favorable Outcomes}}{\text{Possible Outcomes}}$.

The probability that an event will never occur is 0, while the probability that an event will definitely occur is 1.

When multiple events occur, the probability of each event is multiplied to get the probability for all the events.

Concept Check

1. Explain why all probabilities are between 0 and 1.

Practice

Find the probability of the following events.

2. Tossing a regular six-sided die and getting a 6

3. Tossing a coin and having it land on heads or tails

4. Having the board game spinner shown above land on orange or red

5. Having the board game spinner shown above land on blue

Real Numbers
Topic 14.1 Introduction to Real Numbers

Vocabulary
whole numbers • integers • rational number • real numbers • irrational numbers
number line • inequality • less than • greater than

1. _____ are all the rational numbers and all the irrational numbers.

2. A(n) _____ is a statement that shows the relationship between any two real
numbers that are not equal.

Step-by-Step Video Notes
Watch the Step-by-Step Video lesson and complete the examples below.

Example	Notes
1–6. Classify each number as an integer, a rational number, an irrational number, and/or a real number. Circle all that apply. 8 integer rational irrational real $-\dfrac{1}{5}$ integer rational irrational real 1.65 integer rational irrational real $\sqrt{3}$ integer rational irrational real 0.444... integer rational irrational real $\sqrt{9}$ integer rational irrational real	
7. Plot the following real number on a number line. −2 (number line from −3 to 3)	
9. Use a real number to represent the real life situation. A temperature of 131.2°F below zero is recorded in Antarctica.	

Example	Notes
11. Write the following statement using inequality symbols. 2 is less than 7.	
14. Plot the given numbers on a number line, and then replace the question mark with the symbol $<$ or $>$. $-1.5 \ ? \ -5$ $-1.5 \ \boxed{} \ -5$ number line from -5 to 5	

Helpful Hints

The symbol "$<$" is used to represent the phrase "is less than." The symbol "$>$" is used to represent the phrase "is greater than."

Positive numbers are to the right of 0 on a number line. Negative numbers are to the left of 0 on a number line. Zero is neither a positive number nor a negative number.

Concept Check

1. Are there any rational numbers that are not real numbers? Are there any real numbers that are not rational? Explain.

Practice

2. Classify each number as an integer, a rational number, an irrational number, and/or a real number.

   0.3333...

   $\sqrt{16}$

3. Use a real number to represent the situation.

   Kelsey deposits $125 into her savings account.

4. Plot the real number on a number line.

   $2\dfrac{2}{3}$

5. Write the following statements using inequality symbols.

   58 is less than 64.

   -7 is greater than -10.

Real Numbers
Topic 14.2 Inequalities and Absolute Value

Vocabulary

variable   •   inequality phrase   •   less than or equal to
greater than or equal to   •   translate   •   absolute value

1. The _____ of a number is the distance between that number and zero on a number line.

2. A(n) _____ is a letter or symbol that is used to represent an unknown quantity.

Step-by-Step Video Notes
Watch the Step-by-Step Video lesson and complete the examples below.

Example	Notes
1. Translate the phrase into an algebraic inequality. A police officer claimed that a car was traveling at a speed more than 85 miles per hour. (Use the variable s for speed.) Determine the phrase to be translated. Replace the unknown quantity with the given variable. Replace the inequality phrase with the correct inequality symbol. Answer:	
3. Translate the phrase into an algebraic inequality. The owner of a trucking company said that the payload of a truck must be no more than 4500 pounds. (Use the variable p for payload.) Answer:	

Example	Notes
5. Use a number line to find the absolute value of the following number. -1.5 $$\overset{\begin{array}{cccccccc}-3 & & -2 & & -1 & & 0 & & 1 & & 2 & & 3\end{array}}{\longleftrightarrow}$$	

7–10. Find the absolute value of the following numbers.

$|-3.68| = \boxed{}$

$|3.68| = \boxed{}$

$|0| = \boxed{}$

$\left|5\dfrac{9}{10}\right| = \boxed{}$

Helpful Hints

When translating phrases into algebraic inequalities, look for key words or phrases to determine which inequality symbol is most appropriate to use.

The absolute value of a number can be thought of as the numerical part of the number, without regard to its sign.

Concept Check

1. Is the inequality $|-6| \geq |6|$ true? What about $|-6| > |6|$? Explain.

Practice

Translate each phrase into an algebraic inequality.

2. According to the building inspector, the elevator can hold at most 12 people. (Use the variable n for number of people.)

3. You must be at least 54 inches tall to go on the roller coaster. (Use the variable i for the number of inches.)

Find the absolute value of the following numbers.

4. $|-24|$

5. $\left|-6\dfrac{7}{8}\right|$

Real Numbers
Topic 14.3 Adding Real Numbers

Vocabulary
absolute value   •   same sign   •   different signs   •   LCD

1. When adding two numbers with _____, subtract the absolute values, or numerical parts, of the numbers and give the answer the same sign as the larger numerical part.

2. When adding add two numbers of the _____, add the absolute values, or numerical parts, of the numbers and give the answer the same sign as the numbers being added.

Step-by-Step Video Notes
Watch the Step-by-Step Video lesson and complete the examples below.

Example	**Notes**
1. Add $8+4$. Add the numerical parts. Give the answer the sign of the numbers being added. Answer:	
3. Add $-\dfrac{2}{3}+\left(-\dfrac{1}{7}\right)$. Answer:	
4. Add $-8.1+(-2.75)+(-5.03)$. Add from left to right. Answer:	

Example	Notes
5. Add $-6+9$.	
Subtract the numerical parts.	
Give the answer the same sign as the larger numerical part.	
Answer:	

7 & 8. Add.

$$3.7+(-10.5)$$

$$-\frac{3}{4}+\frac{7}{9}$$

Helpful Hints
A number written without a sign is assumed to be positive.

Try to determine the sign of the answer before adding. This makes it easier to avoid sign errors in the final answer.

Concept Check
1. Add $-4.7+(-8.7)+23.5+(-10.1)$.

Practice
Add.
2. $6.27+12.8$

4. $6+(-13)$

3. $-8+(-8)$

5. $-\frac{7}{8}+\frac{2}{3}$

Real Numbers
Topic 14.4 Subtracting Real Numbers

Vocabulary
opposite numbers   •   additive inverses   •   absolute value
difference   •   add the opposite   •   sum

1. To subtract real numbers, _____ of the second number to the first.

2. Two numbers that differ only in sign are _____. They are the same distance from zero on a number line but in opposite directions.

3. Two numbers are _____, or opposites, if they add to equal zero.

Step-by-Step Video Notes
Watch the Step-by-Step Video lesson and complete the examples below.

Example	Notes
1–4. Find the additive inverse, or opposite, of each number. Then, add the additive inverse to the number. 10 −4 7.4 $-\dfrac{5}{7}$	
5. Subtract $10 - 40$. Leave the first number alone. Change the minus sign to a plus sign. Change the sign of the number being subtracted. Add the two numbers using the rules of addition. Answer:	

Example	Notes
7. Subtract $-5-(-9)$.	
Answer:	
9. The temperature in a city is $-8°F$ and the temperature of a neighboring town is $-15°F$. Find the difference in temperature between the two cities. Write as a subtraction problem. Subtract. Answer:	

Helpful Hints

If a is a real number, then the opposite of a, denoted $-(a)$, equals $-a$. The opposite of $-a$ is $-(-a) = a$. Any number plus its opposite is equal to zero; $a + (-a) = 0$.

Subtracting a negative number becomes addition of a positive number.

Concept Check

1. Find the opposite of 6 and subtract it from 20. Then, subtract the result from -15.

Practice

2. Find the additive inverse, or opposite, of the number. Then, add the additive inverse to the number.

    -2.3

3. Subtract $-2.7 - 5.4$.

4. Subtract $\frac{4}{5} - \left(-\frac{3}{5}\right)$.

5. The temperature in a city is $32°F$ and the temperature of another city is $18°F$. Find the difference in temperature between the two cities.

Real Numbers
Topic 14.5 Multiplying Real Numbers

Vocabulary
absolute values   •   product   •   negative factors

1. When multiplying two numbers with the same sign, the _____ will always be positive.

2. When multiplying an odd number of _____, the product is negative.

Step-by-Step Video Notes
Watch the Step-by-Step Video lesson and complete the examples below.

Example	**Notes**
1. Multiply $(7)(-4)$. Multiply the numerical parts of the numbers. $(7)(4) = \boxed{}$ Determine the sign of the answer. $(+)(-) = \boxed{}$ Answer:	
3. Multiply $(-12)(-9)$. Answer:	
4. Multiply $(-3.5)(-2)$. Answer:	

Example	Notes
5–8. Multiply.	

$(-3)(-4) = \boxed{}$

$(-3)(-4)(-5) = \boxed{}$

$(-3)(-4)(-5)(6) = \boxed{}$

$(3)(-4)(-5)(6) = \boxed{}$

Helpful Hints
When multiplying two numbers with the same sign, the product is positive. When multiplying two numbers with different signs, the product is negative.

When multiplying an even number of negative factors, the product is positive. When multiplying an odd number of negative factors, the product is negative.

Concept Check
1. Will the product of $-(-27)(-132)$ be positive or negative? Explain.

Practice
Multiply.

2. $(-0.8)(5)$

3. $\left(-\dfrac{1}{4}\right)\left(-\dfrac{2}{3}\right)$

4. $(-2)(-5)(-3)$

5. $(-2)(-5)(-3)(-8)$

Real Numbers
Topic 14.6 Dividing Real Numbers

Vocabulary
reciprocals   •   division   •   quotient   •   dividend   •   divisor

1. The operation of splitting a quantity or number into equal parts is _____.

2. Two numbers are _____ of each other if their product is 1.

Step-by-Step Video Notes
Watch the Step-by-Step Video lesson and complete the examples below.

Example	**Notes**
1. Find the reciprocal of −3. Write the number as a fraction with 1 as the denominator. Invert the fraction. Answer:	
2. Find the reciprocal of $4\frac{3}{5}$. Write the number as an improper fraction. Answer:	
3–5. Divide. $(-21) \div (-7)$ $3.2 \div 0.8$ $\dfrac{-45}{25}$	

Example	Notes
6. Four friends decide to start a business together. They share a start-up loan of $120,000. If they split the amount of the loan equally between them, how much does each friend owe? Divide the amount of the loan by the number of friends. $$\frac{\$120,000}{4} = \boxed{}$$ Answer:	
7. Divide $32 \div \left(-\frac{8}{3}\right)$. Answer:	

Helpful Hints

To find the reciprocal of a fraction, invert (or flip) the fraction. To find the reciprocal of an integer or mixed number, first write the integer or mixed number as an improper fraction. Note that zero does not have a reciprocal.

To divide signed numbers, divide the absolute values, or numerical parts, and determine the sign of the quotient. If the dividend and the divisor have the same sign, the quotient is positive. If they have different signs, then the quotient is negative.

Concept Check

1. Find the reciprocal of $\frac{3}{4}$. Then, divide 18 by the result.

Practice

2. Find the reciprocal.

   $-\frac{3}{8}$

3. Divide $-56 \div (-7)$.

4. Divide $\left(\frac{-12}{35}\right) \div \left(-\frac{6}{7}\right)$.

5. Divide $12.8 \div (-0.8)$.

Name: _____   Date: _____

Instructor: _____   Section: _____

Real Numbers
Topic 14.7 Properties of Real Numbers

Vocabulary
Identity Property of Addition • Identity Property of Multiplication • Distributive Property
Commutative Property of Addition • Commutative Property of Multiplication
Associative Property of Addition • Associative Property of Multiplication

1. The _____ states that changing the order when multiplying numbers does not change the product.

2. The _____ states that for all real numbers a, b, and c, $a(b+c) = ab + ac$.

3. The _____ states that changing the grouping when adding numbers does not change the sum.

Step-by-Step Video Notes
Watch the Step-by-Step Video lesson and complete the examples below.

Example	Notes
1–5. Determine which property is shown by each equation: the Commutative Property of Addition or Multiplication, the Associative Property of Addition or Multiplication, the Identity Property of Addition, or the Identity Property of Multiplication. $-17 + 1 = 1 + (-17)$ $(-4 + 2) + 3 = -4 + (2 + 3)$ $(6.5)(-2) = (-2)(6.5)$ $-46 + 0 = -46$ $8 \cdot \dfrac{5}{5} = 8$	

Example	Notes
6. Multiply $2(x+3)$. Rewrite using the Distributive Property. Answer:	
8 & 9. Multiply. $-(x-2)$ $-5(3x+2y-6z)$	

Helpful Hints

The Identity Property of Addition states that adding zero to a number does not change the number. The Identity Property of Multiplication states that multiplying a number by one gives us the same number.

The Distributive Property is most useful when it does not contradict the order of operations. That is, it is most used when you cannot perform the operations inside the parentheses.

Concept Check

1. Multiply $2a(4z+y)$. Then, rewrite the result using the Commutative Property of Addition.

Practice

2. Determine which property is shown by the equation.

$$6\left(\frac{1}{2}\cdot 5\right)=\left(6\cdot\frac{1}{2}\right)5$$

4. Multiply $-6(-5x+4)$.

3. Determine which property is shown by the equation.

$$(-2+8)+9=9+(-2+8)$$

5. Multiply $9(-3x+4y-7z)$.

Real Numbers
Topic 14.8 Exponents and the Order of Operations

Vocabulary
base   •   exponent   •   order of operations   •   even power

1. An exponent is used as a shortcut for repeated multiplication. The _____ is the number being multiplied.

Step-by-Step Video Notes
Watch the Step-by-Step Video lesson and complete the examples below.

Example	**Notes**
1. Write in exponential form. $$\left(-\frac{5}{7}\right)\left(-\frac{5}{7}\right)\left(-\frac{5}{7}\right)\left(-\frac{5}{7}\right)\left(-\frac{5}{7}\right)\left(-\frac{5}{7}\right)$$ Answer:	
2–5. Evaluate. $(-4)^2$ $\qquad$ $(-4)^3$ -4^2 $\qquad$ -4^3	
8. Evaluate $-\left(-\dfrac{1}{4}\right)^3$. Answer:	
9 & 10. Evaluate. -94^0 $(-23)^1$	

Example	Notes
11. Evaluate $(-4)^2 - 2(5)^2$. Notice that the parentheses are already simplified. Simplify the exponents first. $(-4)^2 - 2(5)^2 = \square - 2(\square)$ Next, perform the multiplication. Finally, perform the subtraction. Answer:	
12. Evaluate $\dfrac{4^3 + 2(-5)}{\lvert -2 \rvert^3}$. Begin by simplifying the numerator. Now, simplify the denominator. Answer:	
13. Compound interest can be calculated with the formula $A = P(1+r)^t$ where A is the amount of money in the account, P is the principal, r is the interest rate, and t is the number of years. Find the amount of money in an account after 6 years, if the principal is $1500 and the interest rate is 3%. Round your answer to the nearest cent. Answer:	

Helpful Hints

Any non-zero number raised to the 0 power is equal to 1. Any number raised to the 1 power is equal to itself.

When evaluating exponential expressions, be especially careful in determining the sign of your answer when the base is negative.

The acronym for the order of operations is PEMDAS. It stands for Parentheses, Exponents, Multiply and Divide, Add and Subtract. This is the order in which operations are to be performed when simplifying any expression.

When simplifying a fraction, use the order of operations to simplify the numerator and the denominator separately, then perform the division last.

Concept Check

1. What is the sign of the product $(-9+10)^2 \left(\dfrac{-7}{-14} \right)^3$? Explain.

Practice

2. Write in exponential form.

$$(4.8)(4.8)(4.8)(4.8)(4.8)(4.8)(4.8)$$

$$2 \cdot 2 \cdot 2$$

3. Evaluate $\dfrac{-6+(-5)^2}{-(-2)^3+3}$.

4. Evaluate $(-0.4)^2 + (0.3)^2 - (0.25)^1$.

5. Compound interest can be calculated with the formula $A = P(1+r)^t$ where A is the amount of money in the account, P is the principal, r is the interest rate, and t is the number of years. Find the amount of money in an account after 4 years, if the principal is \$2800 and the interest rate is 4%. Round your answer to the nearest cent.

Additional Notes

Name: _____   Date: _____

Instructor: _____   Section: _____

Algebraic Expressions and Solving Linear Equations
Topic 15.1 Evaluating Algebraic Expressions

Vocabulary

term   •   algebraic expression   •   like terms   •   evaluate

1. A(n) _____ is a combination of numbers and variables, operation symbols, and grouping symbols.

2. A(n) _____ is any number, variable, or product of numbers and/or variables.

Step-by-Step Video Notes
Watch the Step-by-Step Video lesson and complete the examples below.

Example	Notes
1–3. Identify the terms in each expression, and identify any like terms. $2x+1$ $3x+4y-7x+5z$ $-2x^2+3x$	
4. Evaluate $3x+6$ for $x=4$. Substitute the given value for the variable. Simplify. Remember to follow the rules for the order of operations. Answer:	
5. Evaluate $5x^2$ for $x=0$ and $x=6$. Answer:	

Example	Notes
6. Evaluate $3y^2 - y$ for $y = -7$. Substitute the given value for the variable using parentheses. Simplify. Answer:	
8. The perimeter of a rectangle can be found by the expression $2(l + w)$, where l is the length and w is the width. Find the perimeter of this rectangle if $l = 5.8$ meters and $w = 8$ meters. 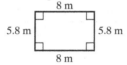 Answer:	

Helpful Hints
The sign in front of the term is considered part of the term. Like terms are terms that have the same variables raised to the same powers.

When substituting negative numbers in algebraic expressions, it is important to use parentheses. This helps to avoid sign errors.

Concept Check
1. Identify the like terms in the expression $4y + 10x - 2y - 5x$. Then, evaluate the expression for $x = 3$ and $y = -2$.

Practice
Identify the terms in each expression, and identify any like terms.

2. $7x + 4y - 3x + 9y - 5$

3. $1.3x^3y + 5.6x + 3.2x^3y - 6.1x$

Evaluate for the given values of l and w.

4. $2(l + w)$ for $l = 4$ and $w = 6.2$

5. $2l + 2w$ for $l = 17$ and $w = 16$

Algebraic Expressions and Solving Linear Equations
Topic 15.2 Simplifying Expressions

Vocabulary
algebraic expression • like terms • simplified • Distributive Property • grouping symbols

1.　An algebraic expression is said to be _____ when none of the terms can be combined.

2.　Many expressions in algebra use _____ such as parentheses ( ), braces { }, or brackets [ ].

Step-by-Step Video Notes
Watch the Step-by-Step Video lesson and complete the examples below.

Example	**Notes**
1–3.　Combine like terms. $$-4a + 3b + 9a = \boxed{}\, a + \boxed{}$$ $$2x^3 + 9x^2 - x + 7$$ $$\frac{4}{3}m + \frac{2}{3}m - m$$	
5.　Combine like terms. $$16x^2 y - 3xy^2 + 5xy - 2x^2 y - 4xy^2$$ Answer:	
6.　Simplify $5 + 4(a + b)$. Use the Distributive Property to remove the parentheses. Answer:	

Example	Notes
7. Simplify $-(4x-3y)$. Answer:	
8. Simplify $2\left[3(9x-8)\right]$. Remove the innermost grouping symbols first. Answer:	

Helpful Hints

Remember that the sign in front of the term is part of the term. When you arrange the terms, the sign in front of the term stays with the term.

To simplify algebraic expressions involving more than one set of grouping symbols, remove the innermost grouping symbols first. Then remove each set of grouping symbols working from the inside to the outside. Finally, combine like terms.

Concept Check

1. Why is the Distributive Property necessary to simplify $4(x+y)$, but not necessary to simplify $4(x+x)$?

Practice

Combine like terms.

2. $5x^3+3x^2-7x^3+4-2x^2$

3. $4-3y+7y^2+9-5y^2-4+8y$

Simplify.

4. $-(8a+3y-2)$

5. $5x+4\left[2x+6(3x-1)\right]$

Algebraic Expressions and Solving Linear Equations
Topic 15.3 Translating Words into Symbols and Equations

Vocabulary
addition   •   multiplication   •   subtraction   •   division   •   equation

1. Phrases like "divided by," "quotient," and "ratio" indicate _____.

2. A(n) _____ is a mathematical statement that two expressions are equal.

Step-by-Step Video Notes
Watch the Step-by-Step Video lesson and complete the examples below.

Example	**Notes**
1–4. Translate into an algebraic expression. Five less than twelve Twelve less than five Fifty divided by one One divided by fifty	
6. Write as an algebraic expression. Use parentheses if necessary. One-half of the sum of a number x and 3 Answer:	
7. Use an expression to describe the measure of each angle. The measure of the second angle of a triangle is double the measure of the first angle, and the third angle is $45°$ less than the measure of the second angle. Answer:	

Example	Notes
8. Translate into an equation. Do not solve. Five more than six times a number is three hundred five. Find the number. Answer:	
10. Translate into an equation. Do not solve. The annual snowfall in Juneau, Alaska is 105.8 inches. This is 20.2 inches less than three times the annual snowfall in Boston, Massachusetts. Find the annual snowfall in Boston. (Snowfall data from www.NOAA.gov) Answer:	

Helpful Hints

The Commutative Property does not work for subtraction or division. This means that the order in which you write your subtraction or division expression matters. For example, "a less than b" translates to $b - a$, not $a - b$. Similarly, "a divided by b" translates to $a \div b$, not $b \div a$.

Concept Check

1. Translate the phrase "a number plus nine, divided by two is seven less than double the number" to an equation. Do not solve.

Practice

Translate into an algebraic expression.

2. Triple the difference of a and b

3. 100 less than the product of 3 and x

Translate into an equation. Do not solve.

4. A number n increased by three is nine. Find the number.

5. The height of a swing set at the park is 10 feet. This is four feet less than seven times the height of the seesaw. Find the height of the seesaw.

Algebraic Expressions and Solving Linear Equations
Topic 15.4 Linear Equations and Solutions

Vocabulary
equation   •   solution   •   variable   •   linear equation in one variable   •   exponent

1. A(n) _____ of an equation is the number(s) that, when substituted for the variable(s), makes the equation true.

2. A(n) _____ is an equation that can be written in the form $Ax + B = C$ where $A, B,$ and C are real numbers and $A \neq 0$.

Step-by-Step Video Notes
Watch the Step-by-Step Video lesson and complete the examples below.

Example	**Notes**
1. Is 2 a solution of the equation $3x - 1 = 5$?	
Substitute 2 for x.	
$3\left(\square\right) - 1 \overset{?}{=} 5$	
Simplify each side of the equation.	
$\square \overset{?}{=} 5$	
Answer:	
2 & 3. Is -1 a solution of the equation $2x + 6 = -1$?	
Is -3 a solution to $7x - 2 = 5$?	
Answer:	

Example	Notes
4–6. Determine if each equation is a linear equation. $2x + 3 = 1$ $2x = 5$ $6x^2 - 3 = 4$	

Helpful Hints

The variable in a linear equation cannot have an exponent greater than 1.

To determine if a given value is a solution of an equation, substitute the given value into the equation. Simplify each side, and if the result is a true statement, that value is a solution.

Concept Check

1. Why is the equation $x(x+4) = 45$ not a linear equation?

Practice

Is 7 a solution of the equation?

2. $42 - 3x = 21$

Determine if each is a linear equation.

4. $5x + 4\frac{1}{3} = -9$

3. $31 = 4x + 5$

5. $11x - 18 = x^2$

Algebraic Expressions and Solving Linear Equations
Topic 15.5 Using the Addition and Multiplication Properties

Vocabulary

Addition Property of Equality   •   equivalent equations   •   solution
solving the equation   •   Multiplication Property of Equality

1. The _____ states that if both sides of an equation are multiplied by the same non-zero number, the solution does not change.

2. The _____ states that if the same number is added to both sides of an equation, both sides of the equation stay equal in value.

3. The process of finding the solution(s) of an equation is called _____.

Step-by-Step Video Notes
Watch the Step-by-Step Video lesson and complete the examples below.

Example	Notes
1. Solve $x + 16.5 = 20.3$ for x. Check your solution. Subtract 16.5 from both sides. Simplify. Check your solution. Solution:	
3. Solve $\dfrac{x}{3} = -15$ for x. Check your solution. Multiply each side of the equation by 3. Simplify. Solution:	

Example	Notes
5. Solve $15 + 2 = 3 + x + 6$ for x. Check your solution. Solution:	
6. Solve $2x - 5x = -12$ for x. Check your solution. Solution:	

Helpful Hints

To solve an equation using the Addition Property of Equality, add or subtract the same number from both sides of the equation to get the variable x on a side of the equation by itself. If a number is being added to x, use subtraction. If a number is being subtracted, use addition.

To solve an equation using the Multiplication Property of Equality, multiply or divide both sides of the equation by the same number to get the variable x on a side of the equation by itself. If x is being multiplied by a number, use division. If x is being divided by a number, use multiplication.

Concept Check

1. Which property of equality should be used to solve the equation $4x - 3x = 1$ for x? Explain.

Practice

Solve for x. Check your solution.

2. $x + 19 = 28$

3. $-25 + 12 = -8 + x + 16$

4. $-32 = -4x$

5. $\dfrac{x}{3} = -9$

Algebraic Expressions and Solving Linear Equations
Topic 15.6 Using the Addition and Multiplication Properties Together

Vocabulary
Addition Property of Equality   •   variable term   •   Multiplication Property of Equality

1. To solve an equation of the form $ax + b = c$, we must use both the _____
   and the _____ together.

Step-by-Step Video Notes
Watch the Step-by-Step Video lesson and complete the examples below.

Example	**Notes**
1. Solve $5x + 3 = 18$ for x to determine how many goals Jenny scored, and then check your solution. Use the Addition Property to subtract 3 from both sides. Use the Multiplication Property to divide both sides by 5. Solution:	
2. Solve $-\dfrac{1}{2}x + 10 = 16$ for x. Use the Addition Property to subtract ☐ from both sides. Use the Multiplication Property to multiply both sides by ☐. Solution:	

Example	Notes
3. Solve $6x - 8 = -2$ for x. Solution:	
4. Solve $4 = -7 + 8x$ for x. Solution:	

Helpful Hints

To evaluate an equation of the form $ax + b = c$, the order of operations tells us to multiply before adding. When trying to solve the equation for x, we must undo this. That is, we must add (or subtract) first, and then multiply (or divide).

Concept Check

1. Which operation would you undo first to solve the equation $-2x + 8 = -14$ for x?

Practice

Solve for x.

2. $7x + 3 = 45$

3. $-22 = 3x - 7$

4. $\frac{1}{6}x + 4 = -8$

5. $4x - 13.2 = 14.8$

Name: _____   Date: _____

Instructor: _____   Section: _____

Solving More Linear Equations and Inequalities
Topic 16.1 Solving Equations with Variables on Both Sides

Vocabulary
solution   •   combining like terms   •   solving the equation

1. The process of finding the solution(s) of an equation is called _____.

Step-by-Step Video Notes
Watch the Step-by-Step Video lesson and complete the examples below.

Example	Notes
1. Solve $9x = 6x + 15$ for x. The goal is to get the variable alone on one side of the equation and numbers on the other side. Subtract $6x$ from both sides. Solution:	
2. Solve $9x + 4 = 7x - 2$ for x. The goal is to get the variable alone on one side of the equation and numbers on the other side. Subtract ☐ x from both sides. Subtract ☐ from both sides. Solution:	

Example	Notes
3. Solve $5x + 26 - 6 = 9x + 12x$ for x. Simplify each side. Get the variable terms on one side. Solution:	
4. Solve $-x + 8 - x = 3x + 10 - 3$ for x. Solution:	

Helpful Hints

Sometimes variable terms and number terms appear on both sides of the equation. If it is necessary, simplify one or both sides of the equation by combining like terms that are on the same side of the equation. Then get the variable terms on one side of the equation and the number terms on the other side.

Concept Check

1. Which term would you add to both sides of the equation $-2x - 8 = 14 - 5x$ so that the variable terms are on one side of the equation with a positive coefficient?

Practice

Solve for x.

2. $4x + 6 = 8x$

4. $7 - 3x + 2 = -9 + 4x - 10$

3. $9x - 22 = 3x - 4$

5. $-12x + 2 + 7x = 1 - 8x + 16$

Solving More Linear Equations and Inequalities
Topic 16.2 Solving Equations with Parentheses

Vocabulary
Distributive Property   •   parentheses   •   equation

1. In order to solve an equation with parentheses, simplify by using the _____
    to remove the parentheses.

Step-by-Step Video Notes
Watch the Step-by-Step Video lesson and complete the examples below.

Example	**Notes**
1. Solve $2(x+5) = -12$ for x. Simplify each side. $2\square + \square = -12$ Notice there is only one variable term. Get the number terms on the other side. Get the variable alone on one side. Solution:	
3. Solve $5(x+1) - 3(x-3) = 17$ for x. Simplify each side. $\square x + \square - \square x + \square = 17$ Get the variable terms on one side. Get the number terms on the other side. Get the variable alone on one side. Solution:	

Example	Notes
5. Solve $3(0.5x-4.2)=0.6(x-12)$ for x.	
Solution:	
6. Solve $2(18x-5)+2=24x-3(12x+8)$ for x.	
Solution:	

Helpful Hints

Recall that the Distributive Property states that for all real numbers $a, b,$ and c, $a(b+c)=ab+ac$. Sometimes an equation has multiple sets of parentheses. If this is the case, apply the Distributive Property as many times as is necessary to remove all sets of parentheses. Also, remember that parentheses can appear inside other parentheses.

Concept Check

1. Can you solve the equation $3(x-7)=12$ without using the Distributive Property?

Practice

Solve for x.

2. $4(x+6)=-8$

3. $7(x-2)-6=x+4$

4. $7(-3x+2)=-8(4x-10)$

5. $0.3x-2(x-1.2)=-0.7(x-3)-3.7$

Solving More Linear Equations and Inequalities
Topic 16.3 Solving Equations with Fractions

Vocabulary
Distributive Property   •   least common denominator   •   equivalent equation

1. To make the process of solving equations with fractions easier, multiply both sides of the equation by the _____ of all the fractions contained in the equation.

Step-by-Step Video Notes
Watch the Step-by-Step Video lesson and complete the examples below.

Example	Notes
1. Solve $\dfrac{1}{4}x - \dfrac{2}{3} = \dfrac{5}{12}x$ for x. Find the LCD of the fractions, then multiply both sides of the equation by the LCD. $\square\left(\dfrac{1}{4}x - \dfrac{2}{3}\right) = \square\left(\dfrac{5}{12}x\right)$ Use the Distributive Property. Solution:	
2. Solve $\dfrac{x}{3} + 3 = \dfrac{x}{5} - \dfrac{1}{3}$ for x. Find the LCD of the fractions, then multiply both sides of the equation by the LCD. $\square\left(\dfrac{x}{3} + 3\right) = \square\left(\dfrac{x}{5} - \dfrac{1}{3}\right)$ Solution:	

Example	Notes
3. Solve $\dfrac{x+5}{7} = \dfrac{x}{4} + \dfrac{1}{2}$ for x. Find the LCD of the fractions, then multiply both sides of the equation by the LCD. Solution:	
4. Solve $0.6x - 1.3 = 4.1$ for x. Solution:	

Helpful Hints

You can also solve an equation containing decimals in a similar way to the fraction equations. You can multiply both sides of the equation by an appropriate value to eliminate the decimal numbers and work only with integer coefficients. If the decimals are tenths, multiply by 10, if the decimals are hundredths or less, then multiply by 100, etc.

Concept Check

1. By what number would you multiply each term in the equation $0.03x - .42 = 1.2$ to work with only integer coefficients?

Practice

Solve for x.

2. $\dfrac{1}{2}x - \dfrac{2}{3} = \dfrac{5}{6}$

4. $\dfrac{x+6}{12} = \dfrac{x}{6} + \dfrac{3}{4}$

3. $\dfrac{7}{8}x - \dfrac{5}{2} = \dfrac{3}{4}x$

5. $3.6 = 4(0.6x - 0.3)$

Name: _____   Date: _____

Instructor: _____   Section: _____

Solving More Linear Equations and Inequalities
Topic 16.4 Solving a Variety of Equations

Vocabulary
identity   •   infinite number of solutions   •   no solution
contradiction   •   solving an equation

1. An equation has _____ if there is no value of x that makes the equation true.

2. An equation has an _____ if the equation is always true, no matter the value of x.

Step-by-Step Video Notes
Watch the Step-by-Step Video lesson and complete the examples below.

Example	Notes
1. Solve $3(6x-4)=4(3x+9)$ for x. Remove the parentheses using the Distributive Property. Solution:	
2. Solve $2(3x+1)=5(x-2)+3$ for x. Solution:	

Example	Notes
3. Solve $\frac{1}{3}(x-2)=\frac{1}{5}(x+4)+2$ for x.	
Solution:	
4. Solve $5(x+3)=2x-8+3x$ for x.	
Solution:	

Helpful Hints
Checking the solution is arguably the most important step of solving an equation. There are situations where possible solutions found may not actually be solutions.

It is important to follow the order of operations when solving equations.

Concept Check
1. Why is an equation such as $2x+9+x=4+3x+5$ called an identity?

Practice
Solve for x.

2. $\frac{1}{4}(4x-12)=\frac{1}{5}(10x+5)$

3. $\frac{1}{3}(x+6)=\frac{1}{6}(x-3)+\frac{2}{3}$

4. $-6+4(x-3)=11x-5-7x$

5. $7(x+3)-6=24-4x-9+11x$

Name: _____   Date: _____

Instructor: _____   Section: _____

Solving More Linear Equations and Inequalities
Topic 16.5 Solving Equations and Formulas for a Variable

Vocabulary
formula   •   Distributive Property   •   least common denominator

1. A _____ is an equation in which variables are used to describe a relationship.

Step-by-Step Video Notes
Watch the Step-by-Step Video lesson and complete the examples below.

Example	Notes
1. Solve $5x + 2 = 17$ and $ax + b = c$ for x. Identify the variable in both equations. Notice that these equations are of the same form, except that every number in the first equation is a variable in the second equation. $5x + 2 = 17 \qquad ax + b = c$ $5x + 2 - \square = 17 - \square \qquad ax + b - \square = c - \square$ $\dfrac{5x}{\square} = \dfrac{\square}{\square} \qquad \dfrac{ax}{\square} = \dfrac{\square}{\square}$ $x = \square \qquad x = \dfrac{\square}{\square}$ Solution:	
2. Solve $d = rt$ for t. Divide both sides of the equation by r. Solution:	

Example	Notes
4. Solve for the specified variable. $a = \dfrac{v}{t}, \ v$ Solution:	
7. Solve for the specified variable. $5x + 3y = 6,$ solve for y Solution:	

Helpful Hints

To solve a formula or an equation for a specified variable, use the same steps for solving a linear equation except treat the specified variable as the only variable in the equation and treat the other variables as if they were numbers.

Sometimes if there are two variables in an equation, you may be asked to solve for one variable *in terms of* the other. For example, "solve $2x - 3y = 6$ for y" indicates that you are to find y in terms of x.

Concept Check

1. What would you divide by to solve $A = bh$ for h?

Practice

Solve for x.

2. $2x + 8y = 12$

3. $y = mx + b$

Solve for the specified variable.

4. $A = \dfrac{1}{2}bh,$ solve for h.

5. $8x - 3y = 12,$ solve for y.

Name: _____   Date: _____
Instructor: _____   Section: _____

Solving More Linear Equations and Inequalities
Topic 16.6 Solving and Graphing Linear Inequalities in One Variable

Vocabulary
inequality   •   linear inequality in one variable   •   solution of an inequality
graph of an inequality   •   solve an inequality   •   non-solutions

1. The _____ is a picture that represents all of the solutions of the inequality.

2. A(n) _____ is a statement that shows the relationship between any two
    real numbers that are not equal.

Step-by-Step Video Notes
Watch the Step-by-Step Video lesson and complete the examples below.

Example	Notes
1 & 2. Graph each inequality on a number line.	

$x > 3$

Use a(n) _____ circle at the boundary
point $x = 3$, because 3 _____ a solution.

Shade all numbers to the _____ side of
the boundary point.

$x \leq -1$

Use a(n) _____ circle at the boundary
point $x = -1$, because -1 _____ a
solution.

Shade all numbers to the _____ side of
the boundary point.

Example	Notes

3. Solve and graph the inequality $5x + 2 < 12$.
 	
    Subtract 2 from both sides of the inequality.
 	
    Divide both sides of the inequality by 5.
 	
    Graph the inequality.
 	
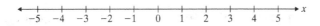	

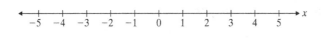

4. Solve and graph the inequality $5 - 4x \geq -7$.

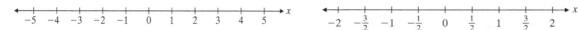

Helpful Hints
It is important to decide if you need an open circle or a closed circle. Remember, for the endpoint, use an open circle for $<$ or $>$ and a closed circle for $\leq$ or $\geq$.

Use the same procedure to solve an inequality that is used to solve an equation, *except* the direction of an inequality must be *reversed* if you *multiply* or *divide* both sides of the inequality *by a negative* number.

Concept Check
1. Would you use an open or closed circle for the boundary point to graph $2x + 3 \leq 7$?

Practice
Graph the inequality on a number line.

2. $x < 2$

Solve and graph the inequality.

4. $8 + 4x > 6$

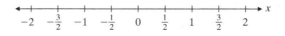

3. $x \geq -\dfrac{1}{2}$

5. $5 - 3x \leq -7$

Solving More Linear Equations and Inequalities
Topic 16.7 Applications of Linear Equations and Inequalities

Vocabulary
variable   •   expression   •   formula   •   equation

1. To write an equation from a word problem, first understand the problem. Then choose a _____ to represent the unknown quantity.

2. To write an equation from a word problem, use a given relationship in the problem or an appropriate _____ to write an equation.

Step-by-Step Video Notes
Watch the Step-by-Step Video lesson and complete the examples below.

Example	Notes
1. Translate the following into an equation and solve. One-third of the length of a race is nineteen miles. Find the length of the race. One-third ⎵ of ⎵ the length of a race ⎵ is ⎵ □ □ l □ ninteen miles. 19 Answer:	
3. Translate the following into an equation and solve. A mother is seven years older than twice the age of her daughter. The sum of their ages is forty-three. Find their ages. Answer:	

Example	Notes
5. Translate the following into an inequality and solve. Heather makes \$14.50 per hour as a lifeguard. Her pay check is calculated by multiplying \$14.50 times the number of hours she worked. How many hours will she have to work to earn more than \$522? $\underline{\$14.50}$ $\underline{\text{times}}$ $\underline{\text{the number of hours}}$ $\underline{\text{is more than}}$ $\underline{\$522}$. Answer:	

Helpful Hints

When choosing a variable, it can help to pick a letter that is associated with the unknown quantity. For example, if you are trying to find a height, choose h as the variable.

Be careful when translating if subtraction or division is involved. The Commutative Property does not work for subtraction or division.

Concept Check

1. Dean wants to buy a new DVD player. His preferred model costs \$204. He earns \$8.50 an hour working part-time. If he works for 20 hours, will he have enough money for the DVD player? Explain.

Practice

Translate the following into an equation and solve.

2. A local pond is 100 feet deep. This is twenty-eight more than three times the depth of another local pond. Find the depth of the second pond.

3. Currently, Bobby's age is five more than three-halves of his sister Jillian's age. If Bobby is 26 years old, how old is Jillian?

Translate the following into an inequality and solve.

4. Marta needs to have more than \$230 dollars to have spending money for her class trip. If she earns \$9.50 per hour working part-time and gets \$40 from her parents for the trip, how many hours does she have to work to have spending money?

Introduction to Graphing Linear Equations
Topic 17.1 The Rectangular Coordinate System

Vocabulary
rectangular coordinate system • x-coordinate • y-coordinate • ordered pair

1. A _____ is made up of a horizontal number line and a vertical number line that intersect to form a right angle. The point where these number lines meet is the origin.

Step-by-Step Video Notes
Watch the Step-by-Step Video lesson and complete the examples below.

Example	**Notes**
1. Plot the points $(4,2)$, $(3,-2)$, $(-3,3)$, and $(-1,-4)$. Label them A, B, C, and D, respectively. 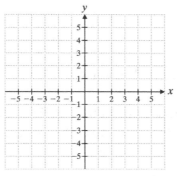	
2. Plot the points on the graph. Identify which quadrant each point lies in. Label them A, B, C, and D, respectively. a. $(4,-5)$ b. $(-5,4)$ c. $(3,0)$ d. $(2,2)$ 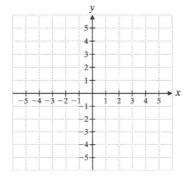	

Example	Notes

3. Find the coordinates of the indicated points.
    Write each point as an ordered pair. Identify
    which quadrant each point lies in.

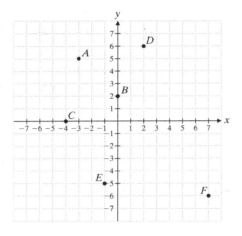

Helpful Hints

The origin is represented by the ordered pair $(0,0)$, and has both x- and y-coordinates of 0.

The quadrants are numbered I, II, III, and IV, starting at the top right and going counter-clockwise. Points that lie on an axis are not considered to be in any quadrant.

Concept Check

1. Give an example of a point in Quadrant IV. Give an example of a point not in a quadrant.

Practice

Plot the given points on the graph. Identify which quadrant each point lies in.

2. $A(-3,4)$ 3. $B(4,-5)$ 4. $C(0,-2)$ 5. $D(4,1)$

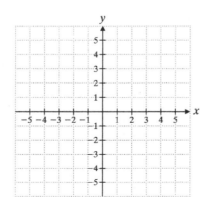

Introduction to Graphing Linear Equations
Topic 17.2 Graphing Linear Equations by Plotting Points

Vocabulary
linear equations in two variables   •   solution to an equation   •   ordered pair

1. A _____ is an equation that can be written in the form $Ax + By = C$ where $A, B,$ and C are real numbers, but A and B are not both zero.

Step-by-Step Video Notes
Watch the Step-by-Step Video lesson and complete the examples below.

Example	Notes
1. Determine whether $(3,5)$ is a solution to the equation $3x + 2y = 19$. Substitute the x- and y-coordinates into the linear equation. Check for a true statement. $3(\square) + 2(\square) \overset{?}{=} 19$ $\square \overset{?}{=} 19$ Answer:	
4. Find three solutions to $2x + y = 13$. Substitute a value for one of the variables. Solve the equation for the other variable. Repeat for the other two solutions. Write the ordered pairs. Answer:	

Example	Notes
6. Find three solutions to $x + y = -4$. Make a table of values to keep your ordered pairs organized. Answer:	

7. Graph the equation $y = 2x + 1$.

   Make a table of values to find three ordered pair solutions. Plot the ordered pairs on a graph. Draw a line through the points.

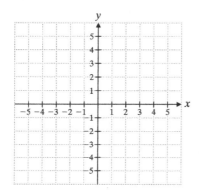 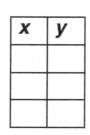

Helpful Hints
To graph a linear equation in two variables, find at least 3 ordered pairs which are solutions of the equation. Plot the ordered pairs, and then draw a line through the points.

Concept Check
1. How many ordered pairs are solutions to the equation $37.5x - 19.2y = 4.8$?

Practice
Is the given point a solution of $2x + 8y = 20$?   Find three solutions to the given equation.

2. $(2, 2)$ 4. $2x - y = 7$

3. $(-6, 4)$ 5. $y = -4x + 1$

Introduction to Graphing Linear Equations
Topic 17.3 Graphing Linear Equations Using Intercepts

Vocabulary
intercept • *x*-intercept • *y*-intercept • origin

1. The point at which a line crosses an axis is called a(n) _____.

2. The _____ is an ordered pair with the coordinates $(a, 0)$, where *a* is a real number.

Step-by-Step Video Notes
Watch the Step-by-Step Video lesson and complete the examples below.

Example	Notes
1. Find the *x*-intercept and *y*-intercept of $3x - 6y = 12$. Find the *x*-intercept by letting $y = 0$ and solving for *x*. Find the *y*-intercept by letting $x = 0$ and solving for *y*. Answer:	
2. Find the *x*-intercept and *y*-intercept of $y = \dfrac{4}{5}x$. Find the *x*-intercept by letting $y = 0$ and solving for *x*. Find the *y*-intercept by letting $x = 0$ and solving for *y*. Answer:	

Example	Notes

3. Graph $2x - y = 4$ using the intercepts.

Make a table of values to find the x-intercept, y-intercept, and another value.

Plot the points on the graph and draw a line through the points.

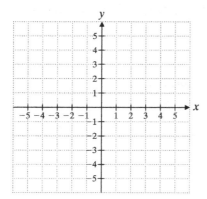

x	y
	0
0	

Helpful Hints
Graphing an equation using the intercepts is exactly the same process as graphing lines by plotting points. In this method, you use two specific points, (the x-intercept and the y-intercept) then plot one more ordered pair and draw a line through the points.

Concept Check
1. Give an example of a linear equation in two variables where both intercepts are the origin, $(0,0)$.

Practice
Find the x- and y-intercepts of each equation.

2. $5x - 4y = 20$

3. $y = \dfrac{1}{2}x + 3$

4. $2 - y = 3x + 2$

5. $y = -4x + 1$

Name: _____   Date: _____

Instructor: _____   Section: _____

Introduction to Graphing Linear Equations
Topic 17.4 Graphing Linear Equations of the Form $x = a$, $y = b$, **and** $y = mx$

Vocabulary

origin   •   horizontal line   •   vertical line

1. If an equation is of the form $y = b$, where b is some real number, then the graph of the equation is a _____.
2. If an equation is of the form $x = a$, where a is some real number, then the graph of the equation is a _____.

Step-by-Step Video Notes
Watch the Step-by-Step Video lesson and complete the examples below.

Example	Notes
1. Graph $6x - 2y = 0$ using three points. Make a table of values. Then, plot the points on a graph and draw a line through the points. 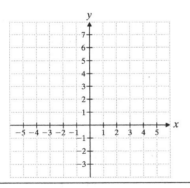	
2. Graph the equation $y = 4$. 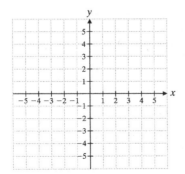	

Example	Notes
5. Graph $3x + 7 = -5$ 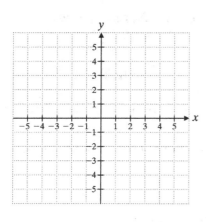	

Helpful Hints

The graph of the vertical line has no y-intercept, unless its equation is $x = 0$.

Similarly, the graph of a horizontal line has no x-intercept, unless its equation is $y = 0$.

Any equation of the form $y = mx$ is neither vertical nor horizontal, and its x- and y-intercepts are the origin, $(0,0)$.

Concept Check

1. The graph of what equation would include all of the points on the x-axis? What are its intercepts?

Practice

Graph each equation.

2. $2x - 4y = 0$

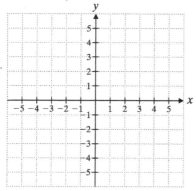

3. $y - 1 = 2$

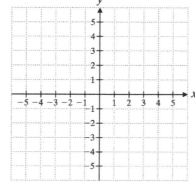

4. $x = -2$

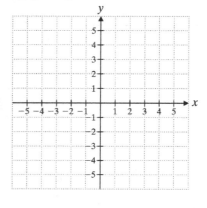

Introduction to Graphing Linear Equations
Topic 17.5 Applications of Graphing Linear Equations

Vocabulary
solution • x-intercept • y-intercept

1. An ordered pair is a(n) _____ to an equation if it results in a true statement when the values for the x-coordinate and the y-coordinate are substituted into the equation.

2. The _____ is an ordered pair with the coordinates $(0,b)$, where b is a real number.

Step-by-Step Video Notes
Watch the Step-by-Step Video lesson and complete the examples below.

Example	Notes
1. Paul received a $75 gift card to an electronics store. Used DVDs cost $3 each, and used video games cost $17 each. The equation that represents buying x DVDs and y games is $3x + 17y = 75$. What does the x-intercept represent in this situation? The x-intercept is the number of _____ Paul can buy if he doesn't buy any _____. Find the x-intercept. Substitute ☐ for y and solve for x. If Paul buys three games, how many DVDs can he buy?	

Example	Notes
3. The amount of money spent annually on food and beverages in restaurants since 2000 can be approximated by the equation $y = 16.8x + 223$, where y is the sales in billions of dollars and x is the number of years since 2000. Graph the equation using the ordered pairs $(0, 223)$, $(6, 323.8)$, $(10, 391)$, and $(14, 458.2)$. Approximate the amount of money spent in restaurants in 2008. 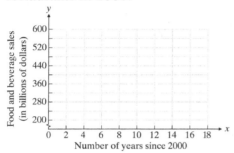	

Helpful Hints

To find a solution to a linear equation in two variables, first substitute a value for one of the variables. Then, solve the equation to find the other variable.

Concept Check

1. Suppose x represents the population of wolves in a forest and y represents the population of rabbits in the same forest. If the relationship between the populations is linear, what do the x- and y-intercepts represent?

Practice

The amount of money spent annually on food and beverages in movie theaters since 2000 can be approximated by the equation $y = 17.5x + 245$, where y is the sales in billions of dollars and x is the number of years since 2000.

2. Graph the equation using four ordered pairs.

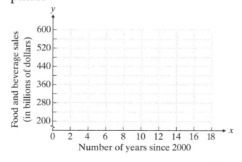

3. Find and interpret the y-intercept.

4. Approximate the amount of money that will be spent in 2016

Slope, Equations of Lines, and Linear Inequalities in Two Variables
Topic 18.1 The Slope of a Line

Vocabulary
rise   •   run   •   slope   •   ordered pair

1. The _____ of a line is the change in horizontal position, or the difference between the x-coordinates of two points on the line.

2. The _____ of a line is the rate of change between any two ordered pair solutions to a linear equation.

Step-by-Step Video Notes
Watch the Step-by-Step Video lesson and complete the examples below.

Example	**Notes**
1. Find the slope. Answer:	
3. Find the slope. Answer:	

Example	Notes
5. Find the slope of the line which contains the points $(1,3)$ and $(5,11)$. Use the slope formula $m = \dfrac{y_2 - y_1}{x_2 - x_1}$. $m = \dfrac{\Box - \Box}{\Box - \Box}$ Answer:	

Helpful Hints

The formula for slope is $\text{Slope} = \dfrac{\text{change in } y\text{-coordinates}}{\text{change in } x\text{-coordinates}} = \dfrac{y_2 - y_1}{x_2 - x_1}$ where $x_1 \neq x_2$.

The variable m is typically used to represent slope.

A line with a positive slope goes up from left to right. A line with a negative slope goes down from left to right. The slope of a horizontal line is zero. The slope of a vertical line is undefined; it can also be said to have no slope. No slope is not the same as zero slope.

Concept Check
1. A line's run is positive and its rise is negative. Is the slope of this line positive or negative?

Practice
Find the slopes of the lines containing the given points.

2. $(2,8)$ and $(0,0)$

3. $(3,1)$ and $(-5,5)$

Find the slopes of the lines.

4. $y = 4$

5. $x = -2.5$

Slope, Equations of Lines, and Linear Inequalities in Two Variables
Topic 18.2 Slope-Intercept Form

Vocabulary
slope-intercept form of a linear equation   • y-intercept   •   Standard Form

1. The _____ that has a slope m and a y-intercept $(0,b)$
   is given by the formula $y = mx + b$.

Step-by-Step Video Notes
Watch the Step-by-Step Video lesson and complete the examples below.

Example	Notes
1. Find the slope and y-intercept of $y = \dfrac{2}{3}x + 5$. Find m. $\dfrac{\square}{\square}$ Find b. $\square$ Answer:	
3. Find the slope and y-intercept of $4x - 3y = 12$. Rewrite the equation in slope-intercept form. Answer:	
6. Find the equation of a line with a slope of 2 and a y-intercept of $\left(0, \dfrac{4}{3}\right)$. Substitute the values for slope and y-intercept into the slope-intercept form. Answer:	

Example	Notes
8. Write the equation of the line shown in the graph. 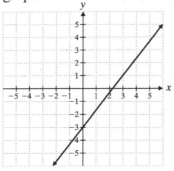 Answer:	

Helpful Hints

You can write an equation in slope-intercept form by solving it for y.

Horizontal lines of the form $y = b$ are simplified from $y = 0x + b$, the slope-intercept form of a line with zero slope. Vertical lines of the form $x = a$ cannot be written in slope-intercept form since they have an undefined slope.

Concept Check

1. How do you know the graph of the equation $y = -3x + 5$ passes through the point $(0, 5)$?

Practice

Find the slope and y-intercept of the given equation.

2. $x - 2y = 6$

3. $y = 3x - 4$

Write the equation of a line with the given slope and y-intercept.

4. $m = 0.5$, $b = -2$

5. $m = -\dfrac{3}{4}$, $b = 6$

Name: _____   Date: _____

Instructor: _____   Section: _____

Slope, Equations of Lines, and Linear Inequalities in Two Variables
Topic 18.3 Graphing Lines Using Slope and *y*-Intercept

Vocabulary
slope   •   *y*-intercept   •   rise   •   run

1. The _____ of a line can be described as "the rise over the run."

Step-by-Step Video Notes
Watch the Step-by-Step Video lesson and complete the examples below.

Example	Notes
1. Find the *y*-intercept of the graph. $b = \boxed{}$ Answer:	
3. Find the slope and the *y*-intercept of the line. Answer:	

Example	Notes

4. Graph $y = \dfrac{1}{2}x - 3$.

The y-intercept is ☐.

The slope is $\dfrac{☐}{☐}$.

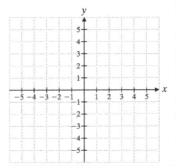

6. Graph $2x + 3y = 6$.

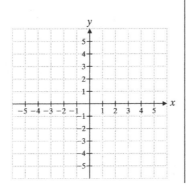

Helpful Hints

If the slope is an integer, its denominator is 1. If x has no coefficient in slope-intercept form, the slope is 1. If there is no constant b in slope-intercept form, the y-intercept is 0.

Concept Check

1. A line has a y-intercept of -2. Explain how to plot another point if the slope is $-\dfrac{2}{3}$.

Practice
Graph each equation.

2. $y = x + 1$

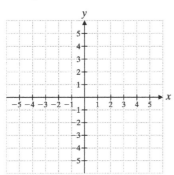

3. $y = \dfrac{1}{3}x - 1$

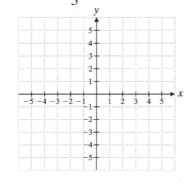

4. $y = -2x + 4$

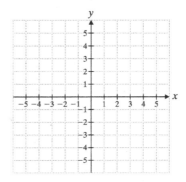

Slope, Equations of Lines, and Linear Inequalities in Two Variables
Topic 18.4 Writing Equations of Lines Using a Point and Slope

Vocabulary
slope-intercept form of a linear equation   •   point-slope form   •   standard form

1. The _____ of an equation of a line whose slope is m and passes through the
   point (x_1, y_1) is given by $y - y_1 = m(x - x_1)$.

Step-by-Step Video Notes
Watch the Step-by-Step Video lesson and complete the examples below.

Example	Notes
1. Use the point-slope form to write the equation of the line with a slope of 2 and passes through $(2,1)$. Write the point-slope equation. Substitute the given values of x_1, y_1, and m into the equation. Simplify. Answer:	
2. Write the equation of the line with a slope of -3 and passes through $(3,9)$. Write your answer in slope-intercept form. Write the point-slope equation. Solve the equation for y. Answer:	

Example	Notes
3. Write the equation of the line with a slope of $\frac{1}{7}$ and passes through $(14,-5)$. Write your answer in slope-intercept form. Answer:	
4. Use the point-slope form to write the equation of the line with a slope of 0 and passes through $(-16,-9)$. Answer:	

Helpful Hints

Point-slope form tells the slope of the line and the coordinates of a point on the line (not necessarily the y-intercept).

Linear equations are usually not left in point-slope form. Most of the time, they will be expressed in slope-intercept form.

Concept Check
1. What is the slope of the graph of the equation $y-32.7=1.5(x-41.3)$?

Practice

Write the equation of the line with the given slope and passes through the given point. Leave your answer in point-slope form.

2. slope of -4 and passes through $(-6,5)$

Write the equation of the line with the given slope and passes through the given point. Write your answer in slope-intercept form.

4. slope of 3 and passes through $(5,7)$

3. slope of $\frac{3}{7}$ and passes through $\left(2,-\frac{8}{9}\right)$

5. slope of $\frac{1}{2}$ and passes through $(-2,-1)$

Slope, Equations of Lines, and Linear Inequalities in Two Variables
Topic 18.5 Writing Equations of Lines Using Two Points

Vocabulary
slope-intercept form   •   point-slope form   •   intercept

1. To write the equation of a line when given two points, use the points to find the slope, then pick one of the points and write the equation in _____.

Step-by-Step Video Notes
Watch the Step-by-Step Video lesson and complete the examples below.

Example	**Notes**
1. Write the equation in slope-intercept form of the line which passes through $(2,5)$ and $(6,3)$. Find the slope m. Substitute the values of x_1, y_1, and m into the point-slope form. Solve for y to write the answer in slope-intercept form. Answer:	
3. Write the equation of the line which passes through $(-5,5)$ and $(0,-6)$. Write your answer in slope-intercept form. Answer:	

Example	Notes
4. Write the equation of the line which passes through $(-8,4)$ and $(-5,0)$. Write your answer in slope-intercept form. Answer:	
5. Write the equation of the line on the graph. Write your answer in slope-intercept form. 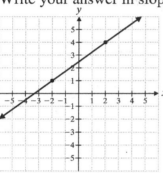 Answer:	

Helpful Hints

Sometimes when given two points to find a line, one of the points is the y-intercept. If this is the case, you can find the slope and then use that point to write the equation in slope-intercept form directly rather than using point-slope form first.

Concept Check

1. Find the slope-intercept form of the line through the points $(-6,5)$ and $(0,3)$ without using point-slope form.

Practice

Write the equation of the line that passes through the given points.

2. $(-2,-5)$ and $(8,5)$

3. $(-8,13)$ and $(-6,5)$

4. $(-11,-4)$ and $(9,8)$

5. $(-2,-2)$ and $(10,4)$

Slope, Equations of Lines, and Linear Inequalities in Two Variables
Topic 18.6 Writing Equations of Parallel and Perpendicular Lines

Vocabulary
parallel lines   •   perpendicular lines   •   negative reciprocals

1. _____ are lines that are always the same distance apart and have no point of intersection.

2. _____ are lines that intersect at a 90° angle.

Step-by-Step Video Notes
Watch the Step-by-Step Video lesson and complete the examples below.

Example	Notes
1. Find the slope of a line parallel to $y = \dfrac{3}{4}x + 2$. Find the slope of the line. Find the slope of a parallel line. Answer:	
2. Find the slope of a line perpendicular to $6x + 2y = 9$. Find the slope of the line. Find the slope of a perpendicular line. Answer:	

Example	Notes
4. Determine if the lines are parallel, perpendicular, or neither. $y = 2x + 3$ $-8x + 4y = -4$ Answer:	
7. Find the equation of a line perpendicular to $y = \dfrac{3}{5}x + 8$ that passes through the point $(3, -1)$. Write the answer in slope-intercept form. Answer:	

Helpful Hints

The symbol for parallel lines is $\parallel$. The symbol for perpendicular lines is $\perp$.

All horizontal lines are parallel to each other and all vertical lines are parallel to each other. Horizontal lines are perpendicular to vertical lines.

Concept Check

1. Is it possible for a set of two lines to be both parallel and perpendicular? Explain.

Practice

Find the slope of a line parallel to and a line perpendicular to the given line.

2. $5x + 3y = 12$

3. $y = 5x + 3.5$

4. Find the slope of a line containing the points $(-2, 3)$ and $(4, 3)$. Then find the slope of a line parallel to this line and the slope of a line perpendicular to this line.

Slope, Equations of Lines, and Linear Inequalities in Two Variables
Topic 18.7 Graphing Linear Inequalities in Two Variables

Vocabulary
linear inequality   •   test point   •   integer solution

1. When graphing a(n) _____, first replace the inequality symbol with an equality symbol. Then graph the line.

Step-by-Step Video Notes
Watch the Step-by-Step Video lesson and complete the examples below.

Example	Notes
1. Graph $4x + 3y \leq 9$. Graph the line. $(\Box,\Box)$ $(\Box,\Box)$ $(\Box,\Box)$ Test a point. $(\Box,\Box)$ $4(\Box) + 3(\Box) \leq 9$ This is a _____ statement. Shade to show all solutions. Answer:	

Example	Notes
2. Graph $3y < -2x$. 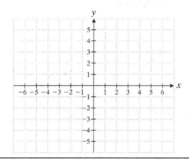	

4. Graph $x > 2$.

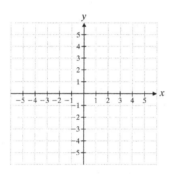

Helpful Hints

The origin $(0,0)$ is usually the most convenient test point to pick. If the point $(0,0)$ is a point on the line, choose another convenient test point.

When the inequality symbol is $\leq$ or $\geq$, use a solid line when graphing the inequality. When the inequality symbol is $<$ or $>$, use a dotted line when graphing the inequality.

Concept Check
1. What test point would you choose to test the linear inequality $y > x$? Explain.

Practice
Graph the given inequality.

2. $y \geq 2x + 3$

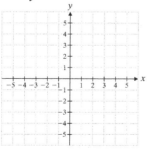

3. $y \leq -2$

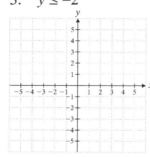

4. $x - y \leq 0$

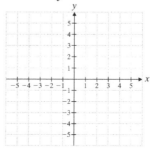

Introduction to Functions
Topic 19.1 Relations and Functions

Vocabulary
relation   •   domain   •   range   •   function   •   ordered pair

1. The second coordinates, or y values, in all of the ordered pairs of a relation make up the
   _____ of the relation.

2. A _____ is a relation for which every x value in the domain has one
   and only one y value.

Step-by-Step Video Notes
Watch the Step-by-Step Video lesson and complete the examples below.

Example	Notes
1. State the domain and range of the relation. $\{(5,7),(9,11),(10,7),(12,14)\}$ State the domain. State the range. Answer:	
2 & 3. Determine whether each relation is a function. $\{(3,9),(4,16),(5,9),(6,36)\}$ $\{(7,8),(9,10),(12,13),(7,14)\}$	

Example	Notes
4–6. Determine whether each relation is a function.	

$y = 3x - 5$

x				
y				

$y = |x|$

x				
y				

$y^2 = x$

x				
y				

Helpful Hints

A relation is any set of ordered pairs. Some relations cannot be expressed by an equation.

The first coordinates, or x values, in all of the ordered pairs of a relation make up the domain of the relation.

Concept Check

1. If you switch the order of the ordered pairs in the relation $\{(-2,4),(-1,1),(1,1),(2,4)\}$, will it still be a function? Explain.

Practice

State the domain and range of the relation.

2. $(8,5)\{(3,9),(9,3),(4,6),(6,4)\}$

3. $y = x^2 - 1$

Determine whether the relation is a function.

4. $x = |y|$

5. $y = x^2 + 4$

Introduction to Functions
Topic 19.2 The Vertical Line Test

Vocabulary

vertical line test   •   *x*-axis   •   ordered pairs

1. The _____ states that if a vertical line can pass along the *x*-axis and cross the graph in at most one place, then the graph represents a function.

Step-by-Step Video Notes
Watch the Step-by-Step Video lesson and complete the examples below.

Example	Notes
1. Determine whether the following is the graph of a function. Answer:	
2. Determine whether the following is the graph of a function. Answer:	

Example	Notes
3. Determine whether the following is the graph of a function. 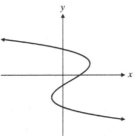 Answer:	
4. Determine whether the following is the graph of a function. 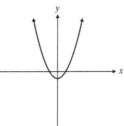 Answer:	

Helpful Hints

A function cannot have two different ordered pairs with the same first coordinate. That is, each value of x must have one and only one value of y.

Concept Check

1. Explain why $y = -2$ is a function, but why $x = -2$ is not a function.

Practice

Determine whether each of the following is the graph of a function.

2.

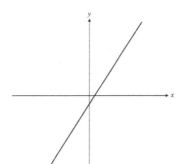

3.

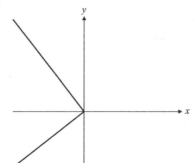

4.

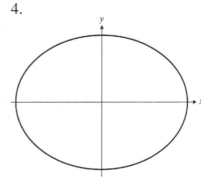

Name: _____

Instructor: _____

Date: _____

Section: _____

Introduction to Functions
Topic 19.3 Function Notation

Vocabulary

function notation   •   domain   •   range

1. If the name of a function is f and the variable is x, the function can be represented by the _____ $f(x)$.

Step-by-Step Video Notes
Watch the Step-by-Step Video lesson and complete the examples below.

Example	Notes
1 & 2. Use function notation to rewrite the following functions using the given function names. $y = 9x - 2$, function name f $y = -16t^2 + 10$, function name h	
3. Determine the domain and range of the function. $$f(x) = -4x + 10$$ Answer:	

Example	Notes

4 & 5. Determine the domain and range of each function.

$$g(x) = x^2 - 4$$

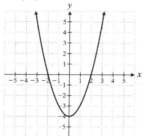

$$h(t) = \sqrt{t}$$

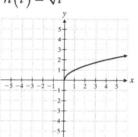

Helpful Hints

Function notation is useful because the variable that makes up the domain is easily identified as the variable inside the parentheses. Read $f(x)$ as " f of x." It does not mean f times x.

Concept Check

1. A function g is defined as $x + y = 3$. Rewrite using the function notation $g(x)$.

Practice

Use function notation to rewrite the functions using the given function names.

2. $y = 3x - 7$, function name f

3. $y = -3x^2 + 5$, function name g

Determine the domain and range of the function based on its graph.

4. $f(x) = \dfrac{4}{3}x - 3$

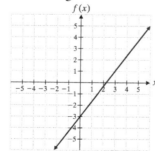

Name: _____ Date: _____

Instructor: _____ Section: _____

Introduction to Functions
Topic 19.4 Evaluating Functions

Vocabulary
function notation • evaluating a function • coordinate

1. When _____ at a certain value, substitute that value for the variable in the expression and simplify.

Step-by-Step Video Notes
Watch the Step-by-Step Video lesson and complete the examples below.

Example	Notes
1 & 2. If $f(x) = x + 8$, find the following. $f(2)$ $f(-6)$	
3–5. If $f(x) = 2x^2 - 4$, find each of the following. $f(5)$ $f(-3)$ $f(0)$	

Example	Notes
6. The approximate length of a man's femur (thigh bone) is given by the function $f(x) = 0.5x - 17$, where x is the height of the man in inches. Find the approximate length of the femur of a man who is 70 inches tall. Substitute 70 for x. Simplify. Answer:	

Helpful Hints

To find a function's value when x is some number a, we write $f(a)$. This point is shown on a graph by the coordinate $(a, f(a))$.

When evaluating a function, it is helpful to place parentheses around the value that is being substituted for x.

When a function defines a real-world relationship, be sure to use the correct units.

Concept Check

1. If $f(x) = (x-2)^2 - 9$, is $f(2) = f(-2)$? Is $f(5) = f(-1)$?

Practice

If $f(x) = 3x - 7$, find the following.   If $h(t) = -16t^2 + 500$, find the following.

2. $f(-2)$ 4. $h(3)$

3. $f(2.6)$ 5. $h(5)$

Name: _____   Date: _____

Instructor: _____   Section: _____

Introduction to Functions
Topic 19.5 Graphing Linear Functions

Vocabulary
function notation • slope-intercept form • linear function • slope • *y*-intercept

1. To graph linear functions in _____, first find the *y*-intercept and the slope.
    Then, plot the *y*-intercept and use the slope to find a second point.

Step-by-Step Video Notes
Watch the Step-by-Step Video lesson and complete the examples below.

Example	Notes
2. Graph $f(x) = \dfrac{2}{3}x - 3$. Identify the slope and *y*-intercept. Graph the function. 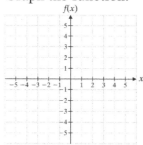	
4. Graph $f(x) = -3x$. 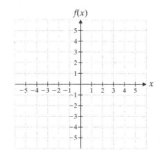	

Example	Notes
5. Graph $f(x) = 4$. 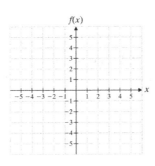	

Helpful Hints

To find a second point using the slope, begin at the y-intercept and move up or down the number of units represented by the numerator of the slope (rise). If the numerator is positive, move up; if it is negative, move down. From that position, move left or right the number of units represented by the denominator (run). If the denominator is positive, move right; if it is negative, move left.

If a function is of the form $f(x) = b$, where b is some real number, then the graph of the function is a horizontal line parallel to the x-axis.

Concept Check

1. Starting at the y-intercept of $f(x) = -4x + 1$, what are the possible second points found using the slope? Explain.

Practice

Graph the following.

2. $f(x) = -2x + 1$ 3. $f(x) = 4x$ 4. $f(x) = -2$

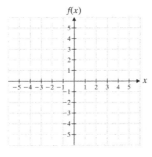   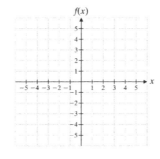   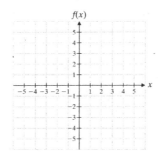

Name: _____   Date: _____

Instructor: _____   Section: _____

Introduction to Functions
Topic 19.6 Applications of Functions

Vocabulary
function   •   relation   •   function notation

1. A _____ is a relation for which every x-value in the domain has one and only one y-value.

Step-by-Step Video Notes
Watch the Step-by-Step Video lesson and complete the examples below.

Example	Notes
1. A turbine wind generator produces P kilowatts of power for a wind speed of w, according to the equation $P = 2.5w^2$. Write the number of kilowatts P as a function of the wind speed w, and find the number of kilowatts generated when the wind speed is 12 mph. $\square(\square) = 2.5w^2$ Answer:	
3. A cell phone company has a text-messaging plan that charges $6.50 per month, plus $0.04 for each text message sent. Write C, the cost per month, as a function of t, the number of text messages sent. What would the monthly cell phone bill be if you sent 2386 text messages in a month? How many text messages were sent if the bill was $149.98?	

Example	Notes

6. When buying hamburger meat for a cookout, Jerry decides that he will need $\frac{1}{3}$ pound of hamburger per person, and he will buy one extra pound of meat to ensure he has enough. This can be represented as the function $f(x) = \frac{1}{3}x + 1$, where x is the number of people coming to the cookout.

Graph this function.

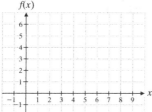

Find the amount of meat Jerry will need to buy if 9 people come to the cookout.

Helpful Hints

If the name of a function is f and the variable is x, the function can be represented by the function notation $f(x)$. Note that this notation does not mean "f multiplied by x."

Concept Check

1. Are there situations that cannot be represented by functions? Explain.

Practice

Jerry decides to plan another cookout. From the last one, he knows that he actually needs $\frac{1}{2}$ pound of hamburger per person, with half a pound extra to ensure he has enough.

2. Write a function f representing the amount of hamburger Jerry will buy if x people attend his cookout.

3. How many people came to Jerry's second cookout if he bought 4 pounds of hamburger?

4. Graph the function.

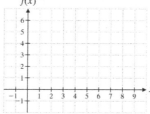

Solving Systems of Linear Equations
Topic 20.1 Introduction to Systems of Linear Equations

Vocabulary
system of linear equations • ordered pair • solution to a system of linear equations

1. A _____ is a set of two or more linear equations containing the same variables.

Step-by-Step Video Notes
Watch the Step-by-Step Video lesson and complete the examples below.

Example	Notes
1. Determine whether $(3,-2)$ is a solution to the following system of equations. $x + 3y = -3$ $4x + 3y = 6$ Answer:	
2. Determine whether $(4,3)$ is a solution to the following system. $7x - 4y = 16$ $5x + 2y = 24$ Answer:	

Example	Notes
4. Determine whether $\left(\dfrac{4}{3}, \dfrac{1}{6}\right)$ is a solution to the following system. $$2y = 1 - \frac{1}{2}x$$ $$3x = 2 + 12y$$	

Answer:

Helpful Hints

If the point (x, y) exists on two lines, then it is a solution to the system of the equations which contains those lines.

Concept Check

1. If $(2,1)$ is the solution to a system of two equations and a third equation is added to the system, is $(2,1)$ still a solution? Explain.

Practice

Determine whether $(4,-1)$ is a solution to the following systems.

2. $x + 2y = 2$
   $5x + 3y = 16$

Determine whether $(0.5, 3)$ is a solution to the following systems.

4. $2x = 3y - 1$
   $\dfrac{1}{3}y = 6x - 2$

3. $3x - 4y = 16$
   $5x + 6y = 14$

5. $0.6x + 0.4y = 1.5$
   $8x + 3y = 12$

Solving Systems of Linear Equations
Topic 20.2 Solving by the Graphing Method

Vocabulary
inconsistent system of equations • dependent system of equations • point of intersection

1. A(n) _____ has no solution. Its graph will be two parallel lines, which do not intersect. There are no ordered pairs in common.

2. A(n) _____ has an infinite number of solutions. This means that the graphs of the two equations will show the same line.

Step-by-Step Video Notes
Watch the Step-by-Step Video lesson and complete the examples below.

Example	Notes
1. Find the solution (the point of intersection of the two lines).	

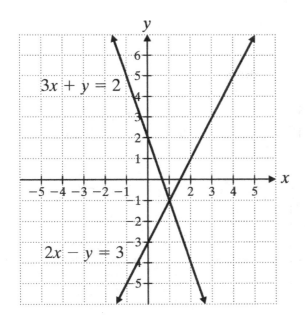

Solution:

Example	Notes
3. Find the solution to the system of equations by graphing. $$y = x - 1$$ $$2x + y = 8$$ Answer:	

Helpful Hints

When you solve a system of linear equations by graphing, graph the equations on the same graph. Find the solution, which is the point of intersection of the lines. Write the solution as an ordered pair. Verify your solution by substituting the ordered pair into both equations.

Concept Check

1. What are the three possible types of solutions to a system of linear equations?

Practice

Find the solution to the system of equations by graphing.

2. $y = x + 1$
    $y = 3x - 1$

3. $x + y = 5$
    $y = \dfrac{1}{2}x - 1$

4. $2x + 3y = 9$
    $y = -2x + 3$

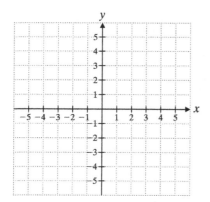 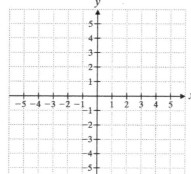 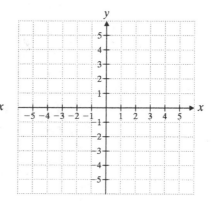

Solving Systems of Linear Equations
Topic 20.3 Solving by the Substitution Method

Vocabulary
substitution method   •   system of equations   •   no solution

1. The _____ involves choosing either equation in a system and solving for either variable and substituting the result into the other equation to solve for the remaining variable.

Step-by-Step Video Notes
Watch the Step-by-Step Video lesson and complete the examples below.

Example	Notes
1. Solve the system of equations by substitution. $$4x + 3y = 50$$ $$y = 2x$$ Substitute $2x$ for y in the first equation. Simplify and solve for x. Use this value for x in one equation to find y. Check your solution in the other equation. Solution:	
2. Solve the system of equations by substitution. $$y = x - 5$$ $$3x - 2y = 17$$ Solution:	

Example	Notes
4. Solve the system of equations by substitution. Solve one equation for one variable. $4x - 24y = 40$ $x - 6y = 10$ Solution:	
5. Solve the system of equations by substitution. $x + 3y = 5$ $4x + 12y = 40$ Solution:	

Helpful Hints
If one of the equations has a variable with a coefficient of 1 or -1, choose that equation to solve. This will almost always make this step easier.

Concept Check
1. Describe a situation where solving a system by substitution would be easier than solving it by graphing.

Practice
Solve the system of equations by substitution.

2. $4x + 3y = 38$
   $y = 5x$

4. $3x + 4y = 8$
   $2x + y = -3$

3. $6x + 5y = 10$
   $x = y + 9$

5. $y = 4x + 7$
   $8x - 2y = 19$

Name: _____   Date: _____

Instructor: _____   Section: _____

Solving Systems of Linear Equations
Topic 20.4 Solving by the Elimination Method

Vocabulary
Elimination Method   •   many solutions   •   no solution

1. When solving systems of equations, you can use the _____ to solve the system if the coefficients of either variable in the equations are opposites.

Step-by-Step Video Notes
Watch the Step-by-Step Video lesson and complete the examples below.

Example	Notes
1. Solve the system of equations. $7x - 3y = 1$ $5x + 3y = 11$ Solution:	
2. Solve the system of equations by the Elimination Method. $3x + 7y = 22$ $2x - 7y = 3$ Check to see if the coefficients of either variable are opposites. Solution:	

Example	Notes
4. Solve the system of equations by the Elimination Method. $4x + 3y = -5$ $7x + 2y = 14$ Solution:	
5. Solve the system of equations by the Elimination Method. $3x + 2y = 18$ $-3x - 2y = 14$ Solution:	

Helpful Hints

An easy way to use elimination is to pick which variable you want to eliminate and multiply each equation by the coefficient of that variable term in the other equation. It may be necessary to also multiply by a negative number to produce opposite variable terms.

Concept Check

1. Describe a situation where solving a system by elimination would be easier than solving by substitution.

Practice

Solve the system by the Elimination Method.

2. $7x + 4y = 8$
   $-7x - 6y = 2$

4. $7x + 5y = 98$
   $8x - 2y = 58$

3. $5x - 6y = 17$
   $2x - 4y = 6$

5. $48x - 33y = 57$
   $-16x + 11y = 19$

Name: _____   Date: _____

Instructor: _____   Section: _____

Solving Systems of Linear Equations
Topic 20.5 Solving a System in Three Variables by the Elimination Method

Vocabulary
system of linear equations   •   ordered triple   •   system in three variables

1. A(n) _____ represents a point in three-dimensional space.

Step-by-Step Video Notes
Watch the Step-by-Step Video lesson and complete the examples below.

Example	**Notes**
1. Solve the system of equations using elimination. $$-2x + 5y + z = 8$$ $$x + 3y - z = 5$$ $$-x + 2y + 3z = 13$$ First, use the first two equations to eliminate z. $$-2x + 5y + z = 8$$ $$x + 3y - z = 5$$ $$\boxed{}x + \boxed{}y \quad = \boxed{}$$ Now, use the second two equations to eliminate z. Solve the resulting system by elimination. Solution:	

Example	Notes
4. Solve the system of equations using elimination. $$-28x - 14y = 42$$ $$2y + 16z = -18$$ $$14x + 6y - 8z = -12$$ Solution:	

Helpful Hints

A system of linear equations in three variables is represented as three planes. If the three planes intersect at a point, there is one solution. If the planes intersect at a common line or if they are the same plane, there are an infinite number of solutions and the system is dependent. If the planes do not all intersect, there is no solution to the system.

When choosing a variable to eliminate, look closely at each equation. Try to eliminate variables that have corresponding coefficients that are multiples. This will simplify the calculations involved in the elimination method.

Concept Check

1. Can a system of three equations in three variables have exactly two solutions? Explain.

Practice

Solve the system of equations using elimination.

2. $-3x + 13y - 10z = -34$
   $2x + 8y - 5z = -19$
   $9x - 5y + 5z = 11$

3. $2x + 5y + 3z = 24$
   $3x + 2y + 5z = -10$
   $7x + 3y + 2z = 2$

4. $-2x + 4y - 6z = 16$
   $5x + 7y + 17z = 1$
   $x - 2y + 3z = 8$

Solving Systems of Linear Equations
Topic 20.6 Applications of Systems of Linear Equations

Vocabulary
system of equations • substitution method • elimination method

1. Sometimes, a(n) _____ can be used to solve application problems.

Step-by-Step Video Notes
Watch the Step-by-Step Video lesson and complete the examples below.

Example	Notes
1. A movie theater sells tickets for $10 and bags of popcorn for $3. In a single Saturday night, the theater had $2375 in sales. The theater owner found that if he raised ticket prices to $11 and the popcorn prices to $4 and sold the same number of tickets and popcorn, the theater would make $2700. How many tickets were sold? How many bags of popcorn were sold? Answer:	
2. A boat travels 20 miles upstream, against the current, in 4 hours. The return trip, 20 miles downstream, with the current, takes only 2 hours. Find the speed of the boat in still water and the speed of the current. Let r = the speed of the boat in still water and c = the speed of the river current. Answer:	

Example	Notes
4. A trucking company has three sizes of trucks. A large truck holds 10 tons of gravel, a medium-sized truck holds 6 tons, and a small truck holds 4 tons. The company wants to use fifteen trucks to haul 104 tons of gravel. To save money on gas, the company's manager wants to use two more large trucks than medium-sized trucks. How many of each type of truck will be used?	

Answer:

Helpful Hints

Remember that once you solve a system relating to an application, it is important to go back and answer the question asked. This requires interpreting the solution to the system.

When you produce a system relating to an application, you can solve it using any method. Choose the method that is easiest for you or that is easiest to use with the particular system that you found.

Concept Check

1. While calculating the answer to Example 2, Tia incorrectly finds that the speed of the current is 0 mph and the speed of the boat is 20 mph. Explain her error.

Practice

2. A plane travels 3150 miles with the wind in 7 hours. The return trip, 3150 miles against the wind, takes 9 hours. What is the speed of the plane?

3. What is the speed of the wind in Practice 2?

4. Bobby spent $8.49 on 11 pieces of candy. The chocolate bars he bought were $1.29, the packs of gum were $1.99, and the sour drops were $0.49. He bought four more sour drops than twice the amount of chocolate bars he bought. How many of each type of candy did Bobby buy?

Introduction to Polynomials and Exponent Rules
Topic 21.1 Introduction to Polynomials

Vocabulary
term • coefficient • like terms • polynomial • descending order • monomial
degree of a term • degree of a polynomial • binomial • trinomial • exponent

1. The _____ is the highest exponent of the base in that term. If there is more than one variable, it is the sum of the exponents of all of the variables in the term.

2. A _____ in the variable x is the sum of a finite number of terms of the form ax^n, where a is any real number and n is a whole number.

Step-by-Step Video Notes
Watch the Step-by-Step Video lesson and complete the examples below.

Example	Notes
2. Find the degree of the term $4ab^2$. The degree of a term is the highest exponent of the base in that term. This term has two variables, _____ and _____. The exponent of _____ is ☐ , and of _____ is ☐ . Answer:	
4. For the polynomial $5x^3 + 8x^2 - 20x - 2$, find the degree of each term. Then find the degree of the polynomial. The degree of the first term is ☐ , the degree of the second term is ☐ , the degree of the third term is ☐ , and the degree of the last term is ☐ since it is a constant. Answer:	

Example	Notes
6. State the degree of the polynomial $5x + 3x^3$, and whether it is a monomial, a binomial, or a trinomial. Answer:	
9. Evaluate the polynomial $P(x) = 2x^2 - 6x + 7$ at $x = -2$. Answer:	

Helpful Hints

You can combine like terms by adding or subtracting the coefficients of the terms.

A polynomial in x is said to be written in descending order if it is written so that the exponents on the variable x decrease from left to right.

The degree of a polynomial is the highest degree of any of its terms. The polynomial 0 is said to have no degree. A polynomial consisting of a constant only is said to have degree 0.

Concept Check

1. Is the polynomial $24x^3 + 8x^2 + 4x + 2$ of greater degree than the monomial $-2x^5$?

Practice

State the degree of the polynomial, and tell if it is a monomial, binomial, or trinomial.

2. $72x^3 + 45x^2 + 27x$

3. $15x^4 y^2 - 22x^3 y$

Evaluate the polynomial at the given value of the variable.

4. $f(x) = 2x^2 + 3x - 9$ for $x = 1.5$

5. $h(t) = -16t^2 + 400$ for $t = 4$

Introduction to Polynomials and Exponent Rules
Topic 21.2 Addition of Polynomials

Vocabulary
adding polynomials   •   coefficients   •   like terms

1. When _____, combine like terms.

Step-by-Step Video Notes
Watch the Step-by-Step Video lesson and complete the examples below.

Example	**Notes**
2. Add. $$\left(5x^2 - 6x - 12\right) + \left(-3x^2 - 9x + 5\right)$$ Remove parentheses and identify like terms. $$5x^2 - 6x - 12 - 3x^2 - 9x + 5$$ Combine like terms. Answer:	
3. Add. $$\left(7x^2 + 8x + 9\right) + \left(13x^2 - 10x + 5\right)$$ Remove parentheses and identify like terms. Combine like terms. We can also use a vertical format to help visualize the addition. $$7x^2 + 8x + 9$$ $$\underline{+13x^2 - 10x + 5}$$ Answer:	

Example	Notes
4. Add. $$(1.2x^3 - 5.6x^2 + 5) + (-3.4x^3 - 1.2x^2 + 4.5x - 7)$$ Answer:	
5. Add. $$\left(\frac{1}{2}x^2 - 6x + \frac{1}{3}\right) + \left(2x - \frac{1}{2} + \frac{1}{5}x^3\right)$$ Answer:	

Helpful Hints

You can add polynomials by combining like terms. While each polynomial may not be given in descending order, arrange the sum in descending order.

If you use a vertical format to add polynomials, be sure each column contains terms of the same degree.

Concept Check

1. How many terms are in the polynomial sum $(4x^3 - 7x^2 + 3x) + (-2x^2 + 9x + 3)$?

Practice

Add.

2. $(-4x^3 + 3x^2 - 2) + (7x^3 - 8x^2 - 3)$

4. $\left(-\frac{1}{2}x^2 - 5x - 6\right) + \left(8x + \frac{3}{4}x^2 + \frac{2}{3}\right)$

3. $(4.4x^3 - 0.13x^2 + 2.11x) + (-0.07x^2 - 1.89x + 6)$

5. $(13x^2 + 2 - 5x) + (13x - 9x^2 + 7)$

Introduction to Polynomials and Exponent Rules
Topic 21.3 Subtraction of Polynomials

Vocabulary
Adding polynomials   •   subtracting polynomials   •   like terms

1. When _____, change the sign of each term in the second polynomial and add the result to the first polynomial by combining like terms.

Step-by-Step Video Notes
Watch the Step-by-Step Video lesson and complete the examples below.

Example	**Notes**
1. Subtract. $(2x+3)-(x-5)$ Change the sign of each term in the second polynomial and add. $2x+3-\square+\square$ Add the polynomials by combining like terms. $2x-\square+3+\square$ Answer:	
2. Subtract. $\left(-2x^3+7x^2-3x-1\right)-\left(-6x^3-9x^2-x+4\right)$ Change the sign of each term in the second polynomial and add. You can also use a vertical format to help visualize the addition. $\begin{array}{r}-2x^3+7x^2-3x-1\\ -\left(-6x^3-9x^2-x+4\right)\end{array} \rightarrow \begin{array}{r}-2x^3+7x^2-3x-1\\ +\boxed{}\end{array}$ Answer:	

Example	Notes
3. Subtract. $$\left(-3x^4 + 5x^2 + 2\right) - \left(6x^3 - 10x^2 + 2x - 1\right)$$ Answer:	
5. Subtract. $$\left(-6x^2y - 3xy + 7xy^2\right) - \left(5x^2y - 8xy - 15x^2y^2\right)$$ Answer:	

Helpful Hints

Use extra care in determining which terms are like terms when polynomials contain more than one variable. Every exponent of every variable in the two terms must be the same if the terms are to be like terms.

If you use a vertical format to subtract polynomials, be sure to change the sign of each term in the second polynomial before adding.

Concept Check

1. How many terms are in the difference $\left(4xy - 7x^2y^2 + 3y\right) - \left(-2x^2y^2 + 9xy + 3\right)$?

Practice

Subtract.

2. $\left(4x^2 + 6x - 9\right) - \left(-5x^2 - 6x + 2\right)$ 　　　　 4. $\left(6.3x^2 - 1.8x + 3.5\right) - \left(3.2x - 0.7x^2 - 4.9\right)$

3. $\left(15x^5 - 8x^3 + 5x - 3\right) - \left(-5x^5 + 12x^4 + 6x^2\right)$ 　 5. $\left(a^4 - 7ab + 3ab^2 - 2b^3\right) - \left(2a^4 + 4ab - 6b^3\right)$

Introduction to Polynomials and Exponent Rules
Topic 21.4 Product Rule for Exponents

Vocabulary
exponential expression   •   product rule for exponents   •   base

1. A variable or a number raised to an exponent is a(n) _____.

2. The _____ states that to multiply two exponential expressions that
   have like bases, keep the base and add the exponents, or $x^a \cdot x^b = x^{a+b}$.

Step-by-Step Video Notes
Watch the Step-by-Step Video lesson and complete the examples below.

Example	Notes
3. Multiply $x^3 \cdot x \cdot x^6$. The base of each factor is $\boxed{}$. Keep this base and add the exponents. Note that even though it is not written, the exponent of the term x is $\boxed{}$. $x^3 \cdot x \cdot x^6 = x^{\left(\boxed{}+\boxed{}+\boxed{}\right)}$ Answer:	
5. Simplify $2^3 \cdot 2^5$, if possible. Write your answer using exponential notation. Answer:	

Example	Notes
7. Multiply $(3a)^2 \cdot (3a)^4$. The base of each factor is $\boxed{}$. Answer:	
10. Multiply $(5ab)\left(-\dfrac{1}{3}a\right)(9b^2)$. Answer :	

Helpful Hints

It is important that you apply the product rule even when the exponent is 1. Every variable that does not have a written exponent is understood to have an exponent of 1.

To multiply exponential expressions with coefficients, first multiply the coefficients, and then multiply the variables with exponents separately.

Concept Check

1. Can you use the product rule for exponents to simplify $p^2 \cdot g^2$? Explain.

Practice

Multiply.

2. $w^{12} \cdot w$

4. $(4xy)\left(-\dfrac{1}{8}x^4y^2\right)(4xy^4)$

3. $(-8x^4)(-5x^3)$

5. $(-6.5pq^5)(-p^3)(2p^4q)$

Introduction to Polynomials and Exponent Rules
Topic 21.5 Power Rule for Exponents

Vocabulary

product rule for exponents   •   quotient raised to a power   •   exponential expression
power rule for exponents   •   product raised to a power   •   exponential notation

1. The _____ states that $\left(x^a\right)^b = x^{ab}$.

2. The rule for a _____ is demonstrated by the equation $\left(\dfrac{x}{y}\right)^a = \dfrac{x^a}{y^a}$,

    if $y \neq 0$.

Step-by-Step Video Notes
Watch the Step-by-Step Video lesson and complete the examples below.

Example	Notes
1–3. Simplify. Write your answer using exponential notation. $\left(x^3\right)^5$ $\left(2^7\right)^3$ $\left(y^2\right)^4$	
4–6. Simplify. $\left(ab\right)^8$ $\left(3x\right)^4$ $\left(-2x^2\right)^3$	

Example	Notes
8. Simplify $\left(\dfrac{3}{w}\right)^4$. Write your answer using exponential notation. Answer:	
9. Simplify $\left(\dfrac{-3x^2z^0}{y^3}\right)^4$. Answer:	

Helpful Hints

If a product in parentheses is raised to a power, the parentheses indicate that each factor inside the parentheses must be raised to that power.

When simplifying expressions by using multiple rules involving exponents, be careful determining the correct sign, especially if there is a negative coefficient.

Concept Check

1. A student simplified $\left(-3x^2\right)^4$ as $-81x^8$. Is this correct? If not, what is the error?

Practice
Simplify.

2. $\left(d^4\right)^5$

3. $\left(-8x^4\right)^2$

4. $\left(\dfrac{p}{2}\right)^5$

5. $\left(-\dfrac{2y^0z^3}{3x^5}\right)^4$

Multiplying Polynomials
Topic 22.1 Multiplying by a Monomial

Vocabulary
Distributive Property   •   product rule for exponents   •   monomial

1. A _____ is a polynomial with exactly one term.

2. To multiply a monomial by a polynomial, use the _____ to multiply the monomial by each term in the polynomial.

Step-by-Step Video Notes
Watch the Step-by-Step Video lesson and complete the examples below.

Example	Notes
1. Multiply $3x^2(5x-2)$. Answer:	
2. Multiply $2x(x^2+3x-1)$. Multiply the monomial by each term in the parentheses. Answer:	

Example	Notes
3. Multiply $-6xy\left(x^3 + 2x^2y - y^2\right)$. Answer:	
5. Multiply $\left(2x^2 - 3x + 8\right)(-7x)$. Use the Commutative Property to write the monomial first. Answer :	

Helpful Hints
Remember with a term such as $7x$ that the exponent on x is 1. When you multiply, add the powers, so $7x \cdot x^2 = 7x^{(1+2)} = 7x^3$.

Multiplication is commutative. The order in which terms are multiplied doesn't matter. A monomial can be after a polynomial, but a monomial is usually written before a polynomial.

Concept Check
1. Which multiplication would you perform first to simplify $(2x)(3y)\left(x^2 + 4y\right)$? Why?

Practice
Multiply.

2. $6x^3\left(-3x^2 + 2x\right)$

4. $-3xy^2\left(x^2 + 8xy - 4y^2\right)$

3. $5x\left(x^2 - 4x - 7\right)$

5. $\left(x^3 - 6x + 12\right)(-3x)$

Name: _____   Date: _____

Instructor: _____   Section: _____

Multiplying Polynomials
Topic 22.2 Multiplying Binomials

Vocabulary

binomial • FOIL method • polynomial

1. The _____ for multiplying two binomials means multiply the first terms, the outer terms, the inner terms, and then the last terms.

2. A _____ is a polynomial with exactly two terms.

Step-by-Step Video Notes
Watch the Step-by-Step Video lesson and complete the examples below.

Example	Notes
1. Multiply $(3x+1)(x+4)$. Multiply each term in the first binomial by each term in the second binomial. $3x\left(\square\right)+3x\left(\square\right)+1\left(\square\right)+1\left(\square\right)$ Combine like terms to simplify the expression. Answer:	
3. Multiply $(4x-9y)(8x-3)$. Answer:	

Example	Notes
4. Multiply $(x+3)(x+5)$ using the FOIL method.	
Multiply the first terms.	
Multiply the outer terms.	
Multiply the inner terms.	
Multiply the last terms.	
Answer:	
6. Multiply $(-2+7x)(-9x+5)$ using the FOIL method.	
Answer:	

Helpful Hints

The FOIL method is just a way to help you remember how to multiply binomials. It is only used to multiply a binomial by a binomial.

Concept Check

1. Would you use the FOIL method when multiplying $(7x^2 - 3x^2)(-4x^3 + 18y)$? Why or why not?

Practice

Multiply.

2. $(x+4)(3x+7)$

3. $(5x-12y)(9x-4)$

Multiply using the FOIL method.

4. $(x^2+3)(x^2+8)$

5. $(4x-6)(-3x-2)$

Multiplying Polynomials
Topic 22.3 Multiplying Polynomials

Vocabulary
polynomial   •   multiplying polynomials   •   Distributive Property

1. When _____, multiply each term in the first polynomial by every term in the second polynomial. Then write the sum of the products in descending order.

Step-by-Step Video Notes
Watch the Step-by-Step Video lesson and complete the examples below.

Example	Notes
1. Multiply $(x+4)(3x^2+x+2)$. Multiply each term in the first polynomial by every term in the second polynomial. Combine like terms. Write the final polynomial in descending order. Answer:	
2. Multiply $(3x-1)(6x^2-5x+8)$. Answer:	

Example	Notes
4. Multiply $(4x^2 + 9x + 7)(2x^2 - 6x - 5)$. Answer:	
5. Multiply $(x+1)(x+2)(x+3)$. Multiply the first two polynomials, then multiply the product by the third polynomial. Answer:	

Helpful Hints

A good way to keep terms organized when multiplying bigger polynomials is to multiply the polynomials vertically. This is like multiplying multi-digit numbers vertically.

You can also multiply three or more polynomials; just multiply them two at a time.

Concept Check

1. Which multiplication would you perform first to simplify $(x+4)(x^2 + 7 - 3x)(x+1)$?
   Why?

Practice

Multiply.

2. $(8x+4)(x^2 + 7x - 6)$

3. $(5x^2 - 9x + 4)(6x - 2)$

4. $(x^2 + 6x + 8)(2x^2 - x - 6)$

5. $(4x - 2)(3x + 7)(x + 5)$

Multiplying Polynomials
Topic 22.4 Multiplying the Sum and Difference of Two Terms

Vocabulary
FOIL method   •   multiplying binomials: a sum and a difference

1. The rule _____ states that $(a+b)(a-b) = a^2 - b^2$, where a and
    b are numbers or algebraic expressions.

Step-by-Step Video Notes
Watch the Step-by-Step Video lesson and complete the examples below.

Example	Notes
1. Multiply $(5x+4)(5x-4)$. Multiply each term in the first polynomial by each term in the second polynomial. Answer:	
2. Multiply $(x+5)(x-5)$. Answer:	

Example	Notes
4 & 5. Multiply. $$\left(2x^2 + 3y\right)\left(2x^2 - 3y\right)$$ $$\left(\frac{1}{4}x - \frac{2}{3}\right)\left(\frac{1}{4}x + \frac{2}{3}\right)$$	

Helpful Hints

Remember that the sum and the difference have to be of the same terms. For example, $(a+b)(a-b) = a^2 - b^2$, but $(a+b)(c-d) \neq ac - bd$.

Remember, when squaring a fraction, you square the numerator and square the denominator.

Concept Check

1. Multiply $(x+y)(x-y)(x^2+y^2)$ without using the FOIL method.

Practice

Multiply.

2. $(x+4)(x-4)$

4. $\left(8x^2 + 11y\right)\left(8x^2 - 11y\right)$

3. $(6x+9)(6x-9)$

5. $\left(\frac{4}{5}n - \frac{2}{7}\right)\left(\frac{4}{5}n + \frac{2}{7}\right)$

Multiplying Polynomials
Topic 22.5 Squaring Binomials

Vocabulary
a binomial squared   •   binomial   •   FOIL method

1. The rule _____ states that $(a+b)^2 = a^2 + 2ab + b^2$ and

   $(a-b)^2 = a^2 - 2ab + b^2$ where a and b are numbers or algebraic expressions.

Step-by-Step Video Notes
Watch the Step-by-Step Video lesson and complete the examples below.

Example	Notes
1. Simplify $(2x-3)^2$. Write the expression as a multiplication problem, then multiply the binomials. Answer:	
2. Simplify $(3x+5)^2$. Answer:	

Example	Notes
4. Simplify $\left(2x^2 + 3y\right)^2$.	
Answer:	
5. Simplify $\left(\dfrac{1}{4}x - \dfrac{2}{3}\right)^2$.	
Answer :	

Helpful Hints

When you square a binomial, the middle term of the product will always be double the product of the terms of the binomial. The product is called a perfect square trinomial.

The Binomial Squared rule is helpful because it is only necessary to find the squares of the two terms in the binomial and twice their product. Using the FOIL method yields the same result, but requires more calculation and simplification.

Concept Check

1. What will be the middle term when you square the binomial $(3x-4)$?

Practice
Simplify.

2. $(x+7)^2$

3. $(5x-9)^2$

4. $\left(14x^3 + 13y\right)^2$

5. $\left(\dfrac{3}{7}y - \dfrac{1}{11}\right)^2$

Dividing Polynomials and More Exponent Rules
Topic 23.1 The Quotient Rule

Vocabulary
quotient rule   •   prime factors method   •   Zero as an Exponent Property

1. The _____ states that for all non-zero numbers x, $\dfrac{x^a}{x^b} = x^{a-b}$.

2. The _____ states that for all non-zero numbers x, $\dfrac{x^a}{x^a} = x^0 = 1$.

Step-by-Step Video Notes
Watch the Step-by-Step Video lesson and complete the examples below.

Example	Notes
2. Simplify $\dfrac{25x^6}{10x^3}$. Write the numerator and denominator in an expanded form, to show the factors. Answer:	
5. Simplify $-\dfrac{16s^6}{32s^2}$. Divide the number parts by the GCF, then use the quotient rule to simplify the variable part. Answer:	

Example	Notes
6. Simplify $\dfrac{7^{10}}{7^4}$ using the quotient rule.	
Answer:	
9. Evaluate $\dfrac{(ax)^4}{(ax)^4}$.	
Answer:	

Helpful Hints

The quotient rule says to divide like bases, keep the base and subtract the exponents.

Any non-zero number raised to an exponent of zero is equal to 1. 0^0 is undefined.

Concept Check

1. Simplify $\dfrac{a^3}{a^0}$ in two different ways.

Practice

Simplify by the prime factors method.

2. $\dfrac{x^8}{x^4}$

3. $-\dfrac{42x^8}{77x^5}$

Simplify using the quotient rule.

4. $\dfrac{x^{49}}{x^7}$

5. $\dfrac{(37np)^9}{(37np)^9}$

Dividing Polynomials and More Exponent Rules
Topic 23.2 Integer Exponents

Vocabulary
negative exponent   •   power rule   •   product rule

1. By definition of a _____, $x^{-n} = \dfrac{1}{x^n}$, if $x \neq 0$.

Step-by-Step Video Notes
Watch the Step-by-Step Video lesson and complete the examples below.

Example	Notes
1–3. Simplify. Write your answers with positive exponents. z^{-6} $\dfrac{x^3}{x^7}$ $\left(x^{-5}\right)\left(x^3\right)$	
4–6. Simplify. Write your answers with positive exponents. 2^{-5} -5^{-2} $(-3)^{-3}$	

Example	Notes
8. Simplify $\dfrac{x^{-4}}{y^{-2}}$. Write your answer with positive exponents. Answer:	
11. Simplify $\dfrac{x^2 y^{-4}}{x^{-5} y^3}$. Write your answer with positive exponents. Answer :	

Helpful Hints

A negative exponent moves the base from one part of the fraction to the other. That is, it moves a numerator to the denominator, and a denominator to the numerator, and the exponent becomes positive.

Concept Check

1. Which is greater, $\left(\dfrac{1}{2}\right)^1$, or $\left(\dfrac{1}{2}\right)^{-1}$?

Practice

Simplify. Write your answers with positive exponents.

2. $2x^{-7}$

Evaluate. Write your answers with positive exponents.

4. $\dfrac{(2x)^{-5}}{y^{-10}}$

3. $\dfrac{x^{-3}}{x^{-5}}$

5. $\dfrac{-3m^{-3}\left(5n^4\right)^{-2}}{m^{-12}n^7}$

Dividing Polynomials and More Exponent Rules
Topic 23.3 Scientific Notation

Vocabulary
Powers of ten   •   scientific notation   •   decimal notation

1. A positive number is written in _____ when it is in the form $a \times 10^n$, where $1 \le a < 10$ and n is an integer.

Step-by-Step Video Notes
Watch the Step-by-Step Video lesson and complete the examples below.

Example	Notes
1. Write $23,400,000$ in scientific notation. Move the decimal point to put one non-zero digit to the left of the decimal point. 2 3 4 0 0 0 0 0 The point was moved ☐ places. Answer:	
3. Write 2.31×10^6 in decimal notation. The positive power of 10 means move the decimal point ☐ places to the _____. Answer:	
6. Write the number in the following application in decimal notation. The volume of a gold atom is 1.695×10^{-23} cubic centimeters. Write this in decimal notation. Answer:	

Example	Notes
7 & 8. Simplify. $\left(8\times10^{16}\right)\left(7\times10^{4}\right)$ $\dfrac{1.5\times10^{7}}{2.5\times10^{2}}$	

Helpful Hints

Write numbers in scientific notation to make a very big or very small number more compact.

To write a number in decimal form, move the decimal point the correct number of places in the appropriate direction. If the power of 10 is positive, the decimal point moves right. If the power of 10 is negative, the decimal point moves left.

Concept Check

1. How many zeros are at the end of the decimal number equivalent to 3.048×10^{7} ?

Practice

Write each number in scientific notation.

2. $24,300,000$

3. $\left(2.5\times10^{-4}\right)\left(4\times10^{6}\right)$

Write in decimal notation.

4. 3.04×10^{9}

5. 2.763×10^{-8}

Dividing Polynomials and More Exponent Rules
Topic 23.4 Dividing a Polynomial by a Monomial

Vocabulary
trinomial   •   monomial   •   polynomial

1. When dividing a polynomial by a monomial, divide each term of the _____ by the _____.

Step-by-Step Video Notes
Watch the Step-by-Step Video lesson and complete the examples below.

Example	Notes
3. Divide $\dfrac{15x^5 + 10x^4 + 25x^3}{5x^2}$. Divide each term in the polynomial by $5x^2$. Write the sum of the results. Answer:	
4. Divide $\dfrac{63x^7 - 35x^6 - 49x^5}{-7x^3}$. Answer:	

Example	Notes
5. Divide $\left(36x^3 - 18x^2 + 9x\right) \div \left(9x\right)$. Answer:	
7. Divide $\dfrac{24x^3 + 16x^2 - 56x}{8x^2}$. Answer :	

Helpful Hints
Split up a fraction into an addition of two or more fractions when trying to simplify the fraction, or with division of a polynomial by a monomial.

If each term does not divide evenly, simplify each individual fraction.

Concept Check
1. Biff states incorrectly that $\left(24x^3 + 8x^2\right) \div \left(8x^2\right) = 3x$. What was his error?

Practice
Divide.

2. $\dfrac{72x^8 + 45x^6 + 27x^4}{9x^2}$

4. $\dfrac{-42x^9 + 24x^7 + 18x^5}{-6x^5}$

3. $\left(150x^4 - 220x^3 + 90x^2\right) \div \left(10x^2\right)$

5. $\dfrac{-33x^8 - 24x^5 + 9x^4}{-9x^4}$

Dividing Polynomials and More Exponent Rules
Topic 23.5 Dividing a Polynomial by a Binomial

Vocabulary
long division   •   polynomial long division   •   quotient

1. When setting up _____, place the terms of the polynomials in descending order. Insert a zero for any missing terms.

Step-by-Step Video Notes
Watch the Step-by-Step Video lesson and complete the examples below.

Example	Notes
1. Divide $(6x^2 + 7x + 2) \div (2x + 1)$. Set the problem up using the long-division symbol. Divide $6x^2$ by $2x$. This is the first term of the answer. $$\begin{array}{r} \boxed{3x} \\ 2x+1{\overline{\smash{\big)}\,6x^2 + 7x + 2}} \\ \underline{6x^2 + 3x} \end{array}$$ Answer:	
2. Divide $(x^3 + 5x^2 + 11x + 4) \div (x + 2)$. $$\begin{array}{r} \boxed{} \\ x+2{\overline{\smash{\big)}\,x^3 + 5x^2 + 11x + 4}} \end{array}$$ Answer:	

Example	Notes
3. Divide $(5x^3 - 24x^2 + 9) \div (5x + 1)$. Answer:	

Helpful Hints

After setting up a polynomial long division problem, divide the first term of the dividend by the first term of the divisor. The result is the first term of the answer. Then proceed as you would with numbers until the degree of the remainder is less than the degree of the divisor.

Concept Check

1. What missing term must you insert in the dividend when dividing $(p^3 - p + 8) \div (p - 4)$?

Practice

Divide.

2. $(6x^2 + 4x - 10) \div (3x + 5)$ 3. $(28x^2 - 15x - 20) \div (4x + 3)$ 4. $(n^3 - n^2 - 4) \div (n - 2)$

Factoring Polynomials
Topic 24.1 Greatest Common Factor

Vocabulary
factor • greatest common factor • factoring a polynomial
prime polynomial • Distributive Property

1. _____ is the process of writing a polynomial as a product of two or more factors.

2. When two or more numbers, variables, or algebraic expressions are multiplied, each is called a _____.

Step-by-Step Video Notes
Watch the Step-by-Step Video lesson and complete the examples below.

Example	Notes
1–3. Find the GCF. $x^3, x^7,$ and x^5 $y, y^4,$ and y^7 x and y^2	
5. Factor out the GCF of $9x^5 + 18x^2 + 3x$. Write each term as the product of the GCF and each term's remaining factors. Answer:	

Example	Notes
6. Factor out the GCF of $8x^3y + 16x^2y^2 - 24x^3y^3$.	

Answer:

---

7 & 8. Factor out the GCF.

$24ab + 12a^2 + 36a^3$

$3x + 7y + 12xy$

Helpful Hints
Factoring a polynomial changes a sum and/or difference of terms into a product of factors.

To find the GCF of two or more terms with coefficients and variables, find the product of the GCF of the coefficients and the GCF of the variable factors.

Concept Check
1. Allison incorrectly states that $3x + 24y - 15z$ is a prime polynomial because the variables in the terms are different. What is her error?

Practice
Factor out the GCF.

2. $72x^8 - 54x^6 + 27x^5$

3. $140x^4 - 210x^3 + 70x^2$

4. $-42x^9y + 24x^7y^2 + 18x^5y^3$

5. $-33x^8y^8 - 24x^5y^5 + 9x^4y^3$

Factoring Polynomials
Topic 24.2 Factoring by Grouping

Vocabulary
common factor   •   factoring by grouping   •   polynomial

1. When _____, collect the terms into two groups so that each group has a common factor. Then factor out the GCF from each group so that the remaining factor in each group is the same.

Step-by-Step Video Notes
Watch the Step-by-Step Video lesson and complete the examples below.

Example	Notes
1. Factor $2x^2 + 3x + 6x + 9$ by grouping. Group terms. Factor within groups. Factor the entire polynomial. Multiply to check your answer. Answer:	
2. Factor $2x^2 + 5x - 4x - 10$ by grouping. Answer:	

Example	Notes
3. Factor $2ax - a - 2bx + b$ by grouping. Answer:	
4. Factor $10x^2 - 8xy + 15x - 12y$ by grouping. Answer:	

Helpful Hints

Sometimes you will need to factor out a negative common factor from the second two terms to obtain two terms that contain the same binomial factor. When factoring out a negative, check your signs carefully.

Concept Check

1. Can you factor Example 1, $2x^2 + 3x + 6x + 9,$ by grouping the terms differently? If so, do you get the same factors?

Practice

Factor by grouping.

2. $6x^2 - 8x + 9x - 12$

3. $6x + 18y + ax + 3ay$

4. $4x + 20y - 3ax - 15ay$

5. $16a^2 - 14ab + 24a - 21b$

Factoring Polynomials

Topic 24.3 Factoring Trinomials of the Form $x^2 + bx + c$

Vocabulary

trinomial   •   FOIL method   •   prime polynomial

1. A _____ is a polynomial that cannot be factored.

Step-by-Step Video Notes

Watch the Step-by-Step Video lesson and complete the examples below.

Example	Notes
1. Factor $x^2 + 7x + 12$. Write the first two terms of the binomial factors. List the possible pairs of factors of 12. Answer:	
2. Factor $x^2 - 8x + 15$. Write the first two terms of the binomial factors. List the possible pairs of factors of 15. Answer:	

Example	Notes
3. Factor $x^2 - 3x - 10$. Answer:	
5. Factor $4x^2 + 40x - 96$. First factor out the GCF. Answer:	

Helpful Hints

Trinomials of the form $x^2 + bx + c$ factor as $(x + \square)(x + \square)$. Trinomials of the form $x^2 - bx + c$ factor as $(x - \square)(x - \square)$, and trinomials of the form $x^2 + bx - c$ factor as $(x + \square)(x - \square)$.

Concept Check

1. To factor a trinomial of the form $x^2 + bx + c$ as the product of two binomials, what must be the product and the sum of the constant terms of the binomial factors?

Practice

Factor.

2. $x^2 + 9x + 18$

4. $3x^2 + 39x + 120$

3. $x^2 - 10x + 24$

5. $x^2 + 13x - 48$

Factoring Polynomials

Topic 24.4 Factoring Trinomials of the Form $ax^2 + bx + c$

Vocabulary

FOIL method   •   reverse FOIL method

1. To factor the trinomial $ax^2 + bx + c$ using the _____, list the different factorizations of ax^2 and c. List the possible factoring combinations until the correct middle term is reached.

Step-by-Step Video Notes

Watch the Step-by-Step Video lesson and complete the examples below.

Example	Notes
1. Factor $2x^2 + 5x + 3$. List the different factorizations of $2x^2$ and 3. List the possible factoring combinations and check the middle term of each combination. Answer:	
2. Factor $4x^2 - 21x + 5$. List the different factorizations of $4x^2$ and 5. Answer:	

Example	Notes
3. Factor $10x^2 - 9x - 9$.	
Answer:	
4. Factor $3x^2 + 5x - 12$.	
Answer:	

Helpful Hints

When factoring trinomials of the form $ax^2 + bx + c$, there can be many combinations of the factors of ax^2 and c. If the trinomial can be factored, only one of these combinations gives the correct middle term.

If c is positive, the signs of both of its factors are the same as the sign of b. If c is negative, then the signs of both of its factors are different.

Concept Check

1. Does the trinomial $8x^2 + 26x + 15$ factor to $(4x + 5)(2x + 3)$? How can you tell without multiplying the binomials completely?

Practice

Factor.

2. $3x^2 + 7x + 2$

4. $9x^2 + 26x + 16$

3. $10x^2 - 37x + 7$

5. $-x^2 - 5x + 24$

Factoring Polynomials

Topic 24.5 Factoring Trinomials by Grouping Numbers (the *ac*-Method)

Vocabulary

grouping   •   the *ac*-method   •   GCF

1. Factoring by _____ is primarily used when the polynomial has four terms and there is no GCF in the four terms.

2. A key part of factoring $ax^2 + bx + c$ by _____ is to find factors of ac that add up to b.

Step-by-Step Video Notes

Watch the Step-by-Step Video lesson and complete the examples below.

Example	Notes
1. Factor $x^2 + 7x + 12$ by the *ac*-method. There is no GCF. Multiply to find *ac*. $ac = \Box \cdot \Box = \Box$ Find the factors of *ac* that add to *b*. Write the trinomial as a four-term polynomial and factor by grouping. Answer:	
2. Factor $8x^2 - 19x + 6$ by the *ac*-method. There is no GCF. Multiply to find *ac*. Find the factors of *ac* that add to *b*. Answer:	

Example	Notes
4. Factor $3x^2 - 6x - 5$ by the ac-method. Answer:	
5. Factor $6x^2 + 11xy + 4y^2$ by the ac-method. Answer:	

Helpful Hints

To factor $ax^2 + bx + c$ by the ac-method, first factor out the GCF, if there is one. Then, multiply a and c. Find the factors of ac that add to b. Finally, rewrite the trinomial and replace bx with the sum of the factors found to make a four-term polynomial and factor by grouping.

If none of the factors of ac add up to b, the trinomial is prime and cannot be factored.

Concept Check

1. Identify the value(s) of b for which the trinomial $3x^2 + bx + 7$ is factorable.

Practice

Factor by the ac-method.

2. $x^2 + x - 6$

3. $3x^2 - 25x + 8$

4. $6x^2 - 26x - 15$

5. $8x^2 + 34xy + 30y^2$

Factoring Polynomials
Topic 24.6 More Factoring of Trinomials

Vocabulary
greatest common factor   •   reverse FOIL method   •   prime polynomial

1. A _____ is a polynomial that cannot be factored.

Step-by-Step Video Notes
Watch the Step-by-Step Video lesson and complete the examples below.

Example	Notes
1. Factor $9x^2 + 3x - 30$. Factor out the GCF. Factor the trinomial inside the parentheses. Answer:	
2. Factor $3 - 10x + 8x^2$. Answer:	

Example	Notes
4. Factor $x^4 - 6x^2 + 9$.	
Answer:	
5. Factor $5x^2 + 7x + 4$.	
Answer:	

Helpful Hints

When factoring a trinomial, always make sure that it is written in descending order.

When factoring trinomials of the form $ax^2 + bx + c$, always look first for a greatest common factor (GCF). This may be the only possible factoring that can be done. Do not forget to include the GCF in your final answer.

When using the reverse FOIL method, creating a table can help keep the factor/sum combinations organized.

Concept Check

1. Find a whole number value of b between 10 and 20 that would make the trinomial $3x^2 + bx + 16$ a prime polynomial.

Practice

Factor.

2. $8x^2 + 8x - 30$

4. $42x + 20x^2 + 2x^3$

3. $42x^3 - 45x^2 + 12x$

5. $7x^3 + 21x^2 - 14x$

More Factoring and Quadratic Equations
Topic 25.1 Special Cases of Factoring

Vocabulary
difference of two squares • perfect square number • binomial squared
perfect square trinomial

1. In a _____, the first and last terms are perfect squares, and the
middle term is twice the products of square roots of the first and last terms.

Step-by-Step Video Notes
Watch the Step-by-Step Video lesson and complete the examples below.

Example	Notes
2–4. Determine if the expression is a difference of two squares. $$x^2 - 16$$ $$x^2 - 7$$ $$4x^2 + 81$$	
5 & 6. Factor. Remember the property $a^2 - b^2 = (a+b)(a-b)$. $$x^2 - 49$$ $$25b^2 - 64$$	

Example	Notes
7 & 8. Factor. $4x^2 - 81y^2$ $-9x^2 + 1$	
9 & 10. Factor the perfect square trinomials completely. $x^2 + 6x + 9$ $9n^2 - 66n + 121$	

Helpful Hints

Other than possibly having a GCF, a sum of two squares will not factor.

If the middle term in a perfect square trinomial is being subtracted, the sign between the terms of the binomial factors will be a minus sign.

Concept Check

1. Can $x^2 - \dfrac{1}{4}$ be factored as a difference of squares? Explain.

Practice

Factor completely.

2. $x^2 - 144$

3. $25x^2 - 81y^2$

4. $16m^2 - 40m + 25$

5. $3x^2 - 42x + 147$

More Factoring and Quadratic Equations
Topic 25.2 Factoring Polynomials

Vocabulary
a difference of two squares   •   perfect square trinomial
greatest common factor   •   reverse FOIL method

1. When factoring a polynomial, always start by looking for a _____.

Step-by-Step Video Notes
Watch the Step-by-Step Video lesson and complete the examples below.

Example	Notes
1. Factor $3k^2 - 48$ completely. Factor out the GCF, if possible. Factor the remaining binomial factor. Answer:	
2. Factor $12n^3 - 12n^2 - 144n$ completely.	

Example	Notes
4. Factor $x^3 + 2x^2 - 9x - 18$ completely.	
5. Factor $4x^3 + 8x^2$ completely.	

Helpful Hints
When factoring polynomials, check for special cases and use different strategies depending on how many terms are in the polynomial. The polynomial will be factored completely when each factor is a prime polynomial.

Concept Check
1. If a polynomial has four terms and no GCF, how should you try to factor?

Practice
Factor completely.

2. $9x^6 - 48x^3 + 64$

3. $-3x^3 + 18x^2 + 48x - 288$

4. $64n^8 - 4$

5. $15x^2 - 23x + 4$

More Factoring and Quadratic Equations
Topic 25.3 Factor the Sum and Difference of Cubes

Vocabulary
perfect cube numbers   •   difference of two cubes   •   sum of two cubes

1. The formula for a _____ is $a^3 + b^3 = (a+b)(a^2 - ab + b^2)$, where a and b are numbers or algebraic expressions.

2. The formula for a _____ is $a^3 - b^3 = (a-b)(a^2 + ab + b^2)$, where a and b are numbers or algebraic expressions.

Step-by-Step Video Notes
Watch the Step-by-Step Video lesson and complete the examples below.

Example	Notes
1–3. Determine if the expression is a sum of two cubes, a difference of two cubes, or neither. $x^3 - 27 = (\square)^3 - (\square)^3$ $x^3 + 75$ $125x^6 + y^{12}$	
4. Factor $x^3 + 125$ completely. Write each term as a perfect cube. $x^3 + 125 = (\square)^3 + (\square)^3$ Write the two factors. $(x + \square)(x^2 - x\square + \square^2)$ Answer:	

Example	Notes
5. Factor $x^3 - 64$ completely. Answer:	
7. Factor $216x^3 - 64y^3$ completely. Make sure to factor out a common factor first. $216x^3 - 64y^3 = \square\left(\square x^3 - \square y^3\right)$ Answer:	

Helpful Hints

The first ten perfect cubes are $1^3 = 1$, $2^3 = 8$, $3^3 = 27$, $4^3 = 64$, $5^3 = 125$, $6^3 = 216$, $7^3 = 343$, $8^3 = 512$, $9^3 = 729$, and $10^3 = 1000$. Like perfect squares, it is a good idea to memorize as many of these as possible so that they are easily recognizable.

Pay special attention to the signs in the factors of a sum or difference of cubes. Remember that for a sum of cubes, $a^3 + b^3 = (a+b)(a^2 - ab + b^2)$, $+ \Rightarrow (+)(-+)$. For a difference of cubes, $a^3 - b^3 = (a-b)(a^2 + ab + b^2)$, $- \Rightarrow (-)(++)$.

Concept Check

1. Explain how to factor $1 - (a-b)^3$.

Practice

2. Determine if the expression is a sum of two cubes, a difference of two cubes, or neither.

   $a^2 - 4$ $343x^3 + 1000$

3. Factor $125y^3 - 729$ completely.

4. Factor $8h^3 + 216p^6$ completely.

5. Factor $1024s^3 + 2000$ completely.

More Factoring and Quadratic Equations
Topic 25.4 Solving Quadratic Equations by Factoring

Vocabulary
Zero Property of Multiplication • quadratic equation • standard form

1. A _____ is an equation of the form $ax^2 + bx + c = 0$ where $a \neq 0$.

Step-by-Step Video Notes
Watch the Step-by-Step Video lesson and complete the examples below.

Example	Notes
1. Solve $x^2 + 4x = 0$ by factoring. Factor completely. Set each factor equal to zero. Solve each equation. Answer:	
3. Solve $10x^2 - x = 2$ by factoring. Answer:	

Example	Notes
5. Solve $4x^2 + 9 = 12x$ by factoring.	
6. Solve $x^2 - 64 = 0$ by factoring.	

Helpful Hints

The highest degree of any term in a quadratic equation is 2.

The solutions to quadratic equations are also called roots or zeros. A quadratic equation has at most two solutions. Sometimes both solutions will be the same number.

Concept Check

1. Write the quadratic equation $25x^2 + 34x = 4x - 9$ in standard form.

Practice

Solve by factoring.

2. $7x^2 - 28x = 0$

3. $3x^2 - 3x - 126 = 0$

4. $24x^2 + 2x = 35$

5. $25x^2 = 80x - 64$

More Factoring and Quadratic Equations
Topic 25.5 Applications

Vocabulary

quadratic equation   •   roots of a quadratic equation

1. When using a _____ to solve for real-world measurements such as time, distance, length, etc., only use the positive roots of the equation

Step-by-Step Video Notes
Watch the Step-by-Step Video lesson and complete the examples below.

Example	Notes
1. The cliff diver jumps from a platform placed on a cliff approximately 144 feet above the surface of the sea. Disregarding air resistance, the height S, in feet, of a cliff diver above the ocean after t seconds is given by the quadratic equation $S = -16t^2 + 144$. How long does it take the diver to reach the water? (Note: The height when he hits the water is 0 feet.) Answer:	
2. A tennis ball is thrown upward with an initial velocity of 8 meters/second. Suppose that the initial height above the ground is 4 meters. Find the height S of the ball after 1 second. At what time t will the ball hit the ground? Remember, the equation is $S = -5t^2 + vt + h$. Answer:	

Example	Notes
4. The length of the base of a rectangle is 7 inches greater than the height. If the total area of the rectangle is 120 square inches, what are the length of the base and height of the rectangle? For a rectangle, $\text{Area} = \text{base} \cdot \text{height}$.	

Answer:

Helpful Hints

The process of solving applications involving quadratic equations is the same as solving quadratic equations in general, with the exception that sometimes there are application specific questions that need to be answered based on the solutions.

Concept Check

1. Write the quadratic equation $3200 = -16t^2 + 480t$ in standard form with a positive coefficient for t^2.

Practice

Solve.

2. An egg is thrown upward with an initial velocity (v) of 9 meters/second. Suppose that the initial height (h) above the ground is 2 meters. At what time t will the egg hit the ground? Use the quadratic equation $S = -5t^2 + vt + h$.

3. A rocket is fired upwards with a velocity (v) of 640 feet per second. Find how many seconds it takes for the rocket to reach a height of 6,400 feet. Use the quadratic equation $S = -16t^2 + 640t$.

4. The length of the base of a rectangle is 4 inches more than twice the height. If the total area of the rectangle is 126 square inches, what are the lengths of the base and height of the rectangle? Remember that for a rectangle, $\text{Area} = \text{base} \cdot \text{height}$.

Name: _____   Date: _____

Instructor: _____   Section: _____

Introduction to Rational Expressions
Topic 26.1 Introduction to Rational Expressions and Functions

Vocabulary
rational number • algebraic expression • rational expression • rational function

1. If a function is defined by a rational expression, it is a(n) _____.

2. A(n) _____ can be written as a fraction of two algebraic expressions, as long as the denominator does not equal 0.

Step-by-Step Video Notes
Watch the Step-by-Step Video lesson and complete the examples below.

Example	Notes
1. Find the domain of $f(x) = \dfrac{24x^2 + 9x}{8x + 3}$. Set the denominator equal to zero and solve. Answer:	
2. Find the domain of $f(x) = \dfrac{x^3 - 9x}{x^2 + 5x + 6}$. Answer:	
3. Find the domain of $f(x) = \dfrac{x^3 + 2x^2 + x + 2}{x^2 + 1}$. Answer:	

Example	Notes
4 & 5. If $f(x) = \dfrac{x^2 + 6x + 8}{x^2 - 16}$, find the function values if $x = -3$ and $x = 4$.	

Helpful Hints

A rational expression is an algebraic fraction. The same rules apply to rational expressions that apply to numerical fractions. Because division by 0 is undefined, the denominator of a rational expression cannot be zero. Thus, the domain of a rational function is the set of all values that do not give a zero in the denominator.

To find the domain of a rational function, set the denominator of the rational expression equal to zero and solve for the variable. The domain is the set of all real numbers except these solutions, which make the denominator zero.

Concept Check

1. Explain why $h\left(-\dfrac{2}{3}\right)$ is undefined for the function $h(x) = \dfrac{4x - 7}{3x^2 - 22x - 16}$. Are there any other undefined function values?

Practice

Find the domain of the rational function and evaluate it at $x = -2$ and $x = 3$.

2. $f(x) = \dfrac{2x^3 - 5x^2 + 3}{2x^2 - 11x - 21}$

Find the domain of the rational function and evaluate it at $x = -5$ and $x = 0$.

4. $h(x) = \dfrac{4x}{x^4 + 5}$

3. $g(x) = \dfrac{1}{4x^2 - 36}$

5. $p(x) = \dfrac{x^2 - 3x + 10}{x^3 - 6x^2 + 9x}$

Introduction to Rational Expressions
Topic 26.2 Simplifying Rational Expressions

Vocabulary
Basic Rule of Fractions   •   simplifying rational expressions   •   common factors

1. The _____ states that for any rational expression $\dfrac{ac}{bc}$ and any

   expressions a, b, and c, (where $b \neq 0$ and $c \neq 0$), $\dfrac{ac}{bc} = \dfrac{a}{b}$.

Step-by-Step Video Notes
Watch the Step-by-Step Video lesson and complete the examples below.

Example	Notes
1. Simplify $\dfrac{21}{39}$. Answer:	
2. Simplify $\dfrac{2x+6}{3x+9}$. Factor the numerator and denominator completely, and divide by common factors. Answer:	

Example	Notes
3. Simplify $\dfrac{x^2+9x+14}{x^2-4}$. Answer:	
5. Simplify $\dfrac{5x^2-45}{45-15x}$. Answer:	

Helpful Hints

Factor the numerator and denominator completely, being aware of special factoring cases like differences of two squares, trinomial squares, monomial factors, and negative factors.

For all polynomials A and B, where $A \neq B$, it is true that $\dfrac{A-B}{B-A} = -1$.

Concept Check

1. What makes the expression $\dfrac{125x^3-9y^2}{9y^2-125x^3}$ easy to simplify? What is the simplified form?

Practice

Simplify.

2. $\dfrac{9x+27}{4x+12}$

4. $\dfrac{x^2+9xy+18y^2}{5x^2+17xy+6y^2}$

3. $\dfrac{5x-4}{12-15x}$

5. $\dfrac{21-4x-x^2}{4x^2-36}$

Introduction to Rational Expressions
Topic 26.3 Multiplying Rational Expressions

Vocabulary
multiplying fractions   •   multiplying rational expressions   •   the quotient rule

1. When multiplying rational expressions, use _____ to simplify the
    variable parts.

Step-by-Step Video Notes
Watch the Step-by-Step Video lesson and complete the examples below.

Example	Notes
1. Multiply $\dfrac{12}{7} \cdot \dfrac{49}{36}$. Answer:	
2. Multiply $\dfrac{2x^2}{3y} \cdot \dfrac{6y}{8x}$. Factor first. Find common factors in the numerators and the denominators. Divide the numerical parts by common factors, and use the quotient rule to simplify the variable parts. Answer:	

Example	Notes
3. Multiply $\dfrac{7x}{22} \cdot \dfrac{11x+33}{7x+21}$.	
Answer:	
5. Multiply $\dfrac{x^2-x-12}{16-x^2} \cdot \dfrac{2x^2+7x-4}{x^2-4x-21}$.	
Answer:	

Helpful Hints

To multiply two rational expressions, find the common factors in the numerators and the denominators. Divide the numerators and denominators by common factors. Then multiply the remaining factors.

Concept Check

1. Are there any common factors to divide in the multiplication $\dfrac{4}{x} \cdot \dfrac{x+4}{4x^2+x}$?

Practice

Multiply.

2. $\dfrac{49x^2}{42y^3} \cdot \dfrac{48y^6}{35x}$

4. $\dfrac{x^4-16}{8x^2+32} \cdot \dfrac{32x^2+24x}{3x^3-4x^2-4x}$

3. $\dfrac{3x-24}{6x+75} \cdot \dfrac{4x+50}{12x-96}$

5. $\dfrac{7-x}{x^2-4x-21} \cdot \dfrac{10-x-x^2}{9x-18}$

Introduction to Rational Expressions
Topic 26.4 Dividing Rational Expressions

Vocabulary
dividing fractions • dividing rational expressions • reciprocal

1. When _____ multiply the first rational expression by the reciprocal of the second rational expression.

Step-by-Step Video Notes
Watch the Step-by-Step Video lesson and complete the examples below.

Example	Notes
2. Divide $\dfrac{-12x^2}{5y} \div \dfrac{18x}{15y}$. Find the reciprocal of the second rational expression and multiply. Divide the numerators and denominators by common factors and then write the remaining factors as one fraction. Answer:	
3. Divide $\dfrac{8x}{14} \div \dfrac{8x-32}{7x-28}$. Answer:	

Example	Notes
4. Divide $\dfrac{x^2+3x-10}{x^2+x-20} \div \dfrac{x^2+4x+3}{x^2-3x-4}$.	
Answer:	

5. Divide $\dfrac{x-5}{3} \div \left(25-x^2\right)$.

Answer:

Helpful Hints

The reciprocal of an integer or an expression is 1 over the integer or the expression. To get the reciprocal of a fraction or a rational expression, invert the fraction or expression.

Remember when multiplying or factoring that $\dfrac{(a-b)}{(b-a)} = -1$.

Concept Check

1. Dividing by $\dfrac{4x}{3+y}$ is the same as multiplying by what rational expression?

Practice
Divide.

2. $\dfrac{8x^4}{45y^3} \div \dfrac{32x^2}{9y^2}$

4. $\dfrac{14x^2+13x+3}{28x^2+5x-3} \div \dfrac{6x^2-7x-5}{12x^2+17x-5}$

3. $\dfrac{11x}{42} \div \dfrac{11x-77}{6x-42}$

5. $\dfrac{3x-5}{6} \div \left(25-9x^2\right)$

Name: _____   Date: _____

Instructor: _____   Section: _____

Adding and Subtracting Rational Expressions
Topic 27.1 Adding Like Rational Expressions

Vocabulary
like rational expressions   •   unlike rational expressions   •   numerator

1. Rational expressions with a common denominator are called _____.

Step-by-Step Video Notes
Watch the Step-by-Step Video lesson and complete the examples below.

Example	Notes
1. Add $\dfrac{5a}{4a+2b}+\dfrac{6a}{4a+2b}$. Add the numerators and keep the denominator the same. Answer:	
2. Add $\dfrac{-7m}{2n}+\dfrac{m}{2n}$. Answer:	
3. Add $\dfrac{-3}{x^2-3x+2}+\dfrac{x+1}{x^2-3x+2}$. Answer:	

Example	Notes
4 & 5. Add. $\dfrac{x+3}{x^2-1}+\dfrac{x}{x^2-1}$ $\dfrac{x}{x+1}+\dfrac{1}{x+1}$	

Helpful Hints

If rational expressions have a common denominator, they can be added in the same way as like fractions. For any rational expressions $\dfrac{a}{b}$ and $\dfrac{c}{b}$, $\dfrac{a}{b}+\dfrac{c}{b}=\dfrac{a+c}{b}$, where $b \neq 0$.

With all calculations with rational expressions, remember to simplify whenever possible by combining like terms, factoring, and dividing by common factors.

Concept Check

1. Are $\dfrac{5}{4x-7}$ and $\dfrac{-3}{(-7)+4x}$ like rational expression?

Practice

Add.

2. $\dfrac{5}{x-8}+\dfrac{3}{x-8}$

4. $\dfrac{3x^2-4x}{3x-7}+\dfrac{5x-14}{3x-7}$

3. $\dfrac{3u}{20q}+\dfrac{2u}{20q}$

5. $\dfrac{4}{x+5}+\dfrac{x+1}{x+5}$

Adding and Subtracting Rational Expressions
Topic 27.2 Subtracting Like Rational Expressions

Vocabulary
rational expression   •   like rational expression   •   common denominator

1. A _____ can be written as a fraction of two algebraic expressions.

Step-by-Step Video Notes
Watch the Step-by-Step Video lesson and complete the examples below.

Example	Notes
1. Subtract $\dfrac{-2a}{3b} - \dfrac{5a}{3b}$. Subtract the numerators and keep the denominator the same. Answer:	
2. Subtract $\dfrac{8x}{2x+3y} - \dfrac{3x}{2x+3y}$. Subtract the numerators and keep the denominator the same. Answer:	
3. Subtract $\dfrac{3x^2+2x}{x^2-1} - \dfrac{10x-5}{x^2-1}$. Answer:	

Example	Notes
4 & 5. Subtract.	

$$\frac{3x}{x-4} - \frac{12}{x-4}$$

$$\frac{3x}{x^2+3x+2} - \frac{2x-8}{x^2+3x+2}$$

Helpful Hints

If rational expressions have the same denominator, they can be subtracted in the same way as fractions. For any rational expressions $\frac{a}{b}$ and $\frac{c}{b}$, $\frac{a}{b} - \frac{c}{b} = \frac{a-c}{b}$, where $b \neq 0$.

The numerator of the fraction being subtracted must be treated as a single quantity. Use parentheses when subtracting and be careful to use the correct signs.

Concept Check

1. Sophie subtracted $\frac{2x}{2x-3} - \frac{-3}{2x-3}$ and got an answer of 1. Is she correct? Explain.

Practice

Subtract.

2. $\frac{7}{5x-2} - \frac{8}{5x-2}$

4. $\frac{-7a}{4b} - \frac{a}{4b}$

3. $\frac{9m}{3m+n} - \frac{5m+7}{3m+n}$

5. $\frac{3x^2+17x}{9x+3} - \frac{x-5}{9x+3}$

Adding and Subtracting Rational Expressions
Topic 27.3 Finding the Least Common Denominator for Rational Expressions

Vocabulary
least common denominator (LCD)   •   like rational expression   •   factor

1. The _____ of two or more rational expressions is the smallest
   expression that each of the denominators will divide into exactly.

Step-by-Step Video Notes
Watch the Step-by-Step Video lesson and complete the examples below.

Example	Notes
1. Find the LCD for $\dfrac{5}{2x-4}$, $\dfrac{6}{3x-6}$. Factor each denominator. List each different factor. List each factor the greatest number of times it occurs in each denominator. Answer:	
2. Find the LCD for $\dfrac{5}{12ab^2c}$, $\dfrac{13}{18a^3bc^4}$. Answer:	

Example	Notes
3–5. Find the LCD. $\dfrac{5}{x+3}, \dfrac{2}{x-4}$ $\dfrac{8}{x^2-5x+4}, \dfrac{12}{x^2+2x-3}$ $\dfrac{x+3}{x^2-6x+9}, \dfrac{10}{2x^2-4x-6}, \dfrac{x}{2}$	

Helpful Hints

If a factor occurs more than once in any one denominator, the LCD will contain that factor repeated the greatest number of times that it occurs in any one denominator.

Be careful when lining up common factors. For example, x and $x-2$ are not common factors, but x and x^2 involve the same factor x, with the highest degree of 2.

Concept Check

1. Is $4x^4$ a common denominator of $\dfrac{5}{2x^3}$ and $\dfrac{y}{2x}$? Is it the LCD of $\dfrac{5}{2x^3}$ and $\dfrac{y}{2x}$? Explain.

Practice

Find the LCD.

2. $\dfrac{5}{9x+24}, \dfrac{11}{21x+56}$

4. $\dfrac{2}{x-3}, \dfrac{7}{x+6}$

3. $\dfrac{13}{30x^2y^3z}, \dfrac{16}{45x^3yz^4}$

5. $\dfrac{x-6}{x^3-4x^2+4x}, \dfrac{9x}{7x^2-21x+14}, \dfrac{4}{x}$

Adding and Subtracting Rational Expressions
Topic 27.4 Adding and Subtracting Unlike Rational Expressions

Vocabulary
equivalent rational expression • unlike rational expressions • like rational expressions

1. To add or subtract _____, rewrite each rational expression as a(n)
_____ whose denominator is the least common denominator.

Step-by-Step Video Notes
Watch the Step-by-Step Video lesson and complete the examples below.

Example	**Notes**
2. Add $\dfrac{5}{xy} + \dfrac{2}{y}$. Answer:	
4. Add $\dfrac{4y}{y^2 + 4y + 3} + \dfrac{2}{y+1}$. Find the LCD. Rewrite each rational expression with the LCD as the denominator. Add the numerators and keep the denominator the same. Simplify if possible. Answer:	

Example	Notes
5. Subtract $\dfrac{3x-4}{x-2}-\dfrac{5x-6}{2x-4}$. Answer:	

6. Subtract $\dfrac{-3}{x^2+8x+15}-\dfrac{1}{2x^2+7x+3}$.

Answer:

Helpful Hints

It can be very easy to make a sign mistake when subtracting two rational expressions. You will find it helpful to place parentheses around the numerator of the second fraction so that you will not forget to subtract the entire numerator.

Remember that you can only add or subtract rational expressions with like denominators.

Concept Check

1. If $a \cdot b = c$, explain the steps you would use to add $\dfrac{x}{a}+\dfrac{x}{c}$.

Practice

Add.

2. $\dfrac{9}{m}+\dfrac{4}{mn}$

3. $\dfrac{3a-b}{a^2-9b^2}+\dfrac{4}{a+3b}$

Subtract.

4. $\dfrac{3x+9}{2x-10}-\dfrac{x-3}{x-5}$

5. $\dfrac{x}{x^2+3x-4}-\dfrac{x}{x^2+6x+8}$

Complex Rational Expressions and Rational Equations
Topic 28.1 Simplifying Complex Rational Expressions by Adding and Subtracting

Vocabulary
complex rational expression   •   least common denominator

1. A _____ (also called a complex fraction) is a rational expression that
    contains a fraction in the numerator, in the denominator, or both.

Step-by-Step Video Notes
Watch the Step-by-Step Video lesson and complete the examples below.

Example	**Notes**
1. Simplify. $$\dfrac{\dfrac{1}{x}}{\dfrac{2}{y^2}+\dfrac{1}{y}}$$ Simplify the denominator into a single fraction. Divide the fraction in the numerator by the fraction in the denominator. Answer:	
2. Simplify. $$\dfrac{\dfrac{1}{x}+\dfrac{1}{y}}{\dfrac{3}{x}-\dfrac{2}{y}}$$ Answer:	

Example	Notes
4. Simplify. $$\dfrac{\dfrac{3}{a+b}-\dfrac{3}{a-b}}{\dfrac{5}{a^2-b^2}}$$ Answer:	

Helpful Hints

The fraction bar in a complex fraction is both a grouping symbol and a symbol for division.

To simplify a complex rational expression by adding and subtracting, simplify the numerator and the denominator into single fractions as necessary. Divide the fraction in the numerator by the fraction in the denominator.

Concept Check

1. Write a multiplication problem that is equivalent to the complex fraction $\dfrac{\dfrac{4-x}{y}}{3+x}$.

Practice
Simplify.

2. $\dfrac{\dfrac{1}{a}+\dfrac{1}{a^2}}{\dfrac{2}{b^2}}$

4. $\dfrac{\dfrac{x}{x^2+4x+3}+\dfrac{2}{x+1}}{x+1}$

3. $\dfrac{\dfrac{1}{x}+\dfrac{1}{y}}{\dfrac{x}{2}-\dfrac{5}{y}}$

5. $\dfrac{\dfrac{6}{x^2-y^2}}{\dfrac{1}{x-y}+\dfrac{3}{x+y}}$

Complex Rational Expressions and Rational Equations
Topic 28.2 Simplifying Complex Rational Expressions by Multiplying by the LCD

Vocabulary
complex rational expressions   •   least common denominator (LCD)

1. One way to simplify a complex fraction is to find the _____ of each denominator in the complex fraction and multiply it by both the numerator and the denominator of the complex fraction.

Step-by-Step Video Notes
Watch the Step-by-Step Video lesson and complete the examples below.

Example	Notes
1. Simplify by multiplying by the LCD. $$\dfrac{\dfrac{3}{x}}{\dfrac{2}{x^2}+\dfrac{5}{x}}$$ Determine the LCD. Multiply the numerator and denominator by the LCD. Answer:	
2. Simplify by multiplying by the LCD. $$\dfrac{\dfrac{5}{ab^2}-\dfrac{2}{ab}}{3-\dfrac{5}{2a^2b}}$$ Answer:	

Example	Notes
3. Simplify by multiplying by the LCD. $$\dfrac{\dfrac{3}{a+b}-\dfrac{3}{a-b}}{\dfrac{5}{a^2-b^2}}$$	

Answer:

Helpful Hints
To simplify a complex rational expression by multiplying by the LCD, you will often have to factor the denominators of the fractions in the expression to determine the LCD.

Concept Check
1. When is it easier to simplify a complex fraction by multiplying by the LCD?

Practice
Simplify by multiplying by the LCD.

2. $\dfrac{\dfrac{3}{a}+\dfrac{2}{b}}{\dfrac{5}{ab}}$

3. $\dfrac{\dfrac{3}{4x^2}-\dfrac{2}{y}}{\dfrac{7}{xy}-6}$

4. $\dfrac{\dfrac{8}{x^2-y^2}}{\dfrac{3}{x+y}+\dfrac{4}{x-y}}$

5. $\dfrac{\dfrac{2x}{x+3}+\dfrac{12}{x^2+8x+15}}{\dfrac{3}{x+5}}$

Complex Rational Expressions and Rational Equations
Topic 28.3 Solving Rational Equations

Vocabulary

rational expression   •   extraneous solution   •   apparent solution

1. A(n) _____ is an apparent solution that does not satisfy the original
    equation.

Step-by-Step Video Notes
Watch the Step-by-Step Video lesson and complete the examples below.

Example	Notes
1. Solve for x. $$\frac{5}{x} + \frac{2}{3} = -\frac{3}{x}$$ Multiply each term of the equation by the LCD and solve the resulting equation. Answer:	
2. Solve for x. $$\frac{6}{x+3} = \frac{3}{x}$$ Answer:	

Example	Notes
3. Solve for x. $$\frac{3}{x+5}-1=\frac{4-x}{2x+10}$$	
Answer:	
4. Solve for y. $$\frac{y}{y-2}-4=\frac{2}{y-2}$$	
Answer:	

Helpful Hints

In the case where a value makes a denominator in the equation zero, it is not a solution to the equation and therefore is not included in the domain.

Concept Check

1. Is 3 an extraneous solution to the equation $\dfrac{x}{x-3}+\dfrac{4}{5}=\dfrac{3}{x-3}$?

Practice

Solve for x.

2. $\dfrac{8}{x}+\dfrac{1}{2}=-\dfrac{2}{x}$

4. $\dfrac{2}{x-5}+1=\dfrac{3x-5}{4x-20}$

3. $\dfrac{9}{7x-4}=\dfrac{3}{2x}$

5. $\dfrac{2x}{x-3}=\dfrac{6}{x-3}+3$

Complex Rational Expressions and Rational Equations
Topic 28.4 Applications of Rational Equations: Solving Formulas for a Variable

Vocabulary
formula   •   LCD   •   rational equation

1. When solving a _____ for a variable, the solution is of the form
   variable = expression or expression = variable.

Step-by-Step Video Notes
Watch the Step-by-Step Video lesson and complete the examples below.

Example	Notes
1. The following formula is used in the study of light passing through a lens. It relates the focal length f of the lens to the distance a of an object from the lens and the distance b of the image from the lens. Solve for the variable a. $$\frac{1}{f} = \frac{1}{a} + \frac{1}{b}$$ Find the LCD of all the denominators. Answer:	
2. The gravitational force F between two masses m_1 and m_2 a distance d apart is represented by the following formula. In this equation, G is the gravitational constant. Solve for m_2. $$F = \frac{Gm_1m_2}{d^2}$$ Answer:	

Example	Notes
3. The slope of a line that passes through the points (x_1, y_1) and (x_2, y_2) is shown below. Solve for x_1. $$m = \frac{y_2 - y_1}{x_2 - x_1}$$ Answer:	

Helpful Hints

To solve an equation (or formula) containing rational expressions, first find the LCD of all the denominators. Then, multiply each term of the equation by the LCD. Simplify, then solve the resulting equation for the specified variable.

The only difference between solving an equation and solving a formula is that a formula contains more than one variable. Be sure to remember which variable you are solving for.

Concept Check

1. When solving for x_1 in Example 3, Ben got $x_1 = mx_2 - y_2 + y_2$. Explain his error.

Practice

Solve the equation for the indicated variable.

2. $m = \dfrac{y_2 - y_1}{x_2 - x_1}$, y_1

4. $\dfrac{P}{L + W} = 2$, W

3. $\dfrac{hb_1 + hb_2}{A} = 2$, h

5. $\dfrac{A}{\pi r_1} = r_2$, r_1

Complex Rational Expressions and Rational Equations
Topic 28.5 Applications of Rational Equations: Work Problems

Vocabulary
rate at completing a task   •   work problems   •   rate   •   LCD

1. Problems in which two or more people (or machines) are working together to complete a certain task are known as _____.

2. The property for the _____ states that if a task can be completed in t hours, then $\frac{1}{t}$ of the task can be completed in 1 hour.

Step-by-Step Video Notes
Watch the Step-by-Step Video lesson and complete the examples below.

Example	Notes
1–3. If Alisha can complete the following tasks in the following times, how much of a task can she complete in an hour? In other words, what is her rate at completing the task? Mow the lawn: 2 hours Paint a room: 3 hours Complete math homework: 1.5 hours	
4. Robert can paint the kitchen in 2 hours. Susan can paint the kitchen in 3 hours. How long will it take Robert and Susan to paint the kitchen if they work together? $\dfrac{1}{\square} + \dfrac{1}{\square} = \dfrac{1}{\square}$ Answer:	

Example	Notes
5. Miguel can rake up the leaves in his yard in 6 hours. His neighbor's son can clean up the leaves in Miguel's yard using his leaf blower. If it takes them 1.5 hours to clear the yard together, how long would it take Miguel's neighbor's son to do the job alone?	

Answer:

Helpful Hints
It can be helpful to set up a table to organize the information needed for work problems.

Remember to multiply by the LCD as the first step when solving these types of problems. It can make the calculations much simpler.

Concept Check
1. Refer to Example 5, if Miguel worked at the same rate, but his neighbor's son could actually get the job done by himself in 1.5 hours, how long would it take them to complete the job together?

Practice
2. Jon can paint his mother's house in 24 hours. Chuck, a painter, can paint the house in 12 hours. How long would it take to complete the job if Jon and Chuck worked together?

4. A rain barrel can be filled by a rain gutter during a light rain storm in 6 hours. During the last storm, the barrel sprung a leak that causes the barrel to empty in 10 hours. How long does it take to fill the rain barrel now? (Use a negative number to represent the water lost.)

3. Refer to Practice 2, if Chuck worked with Jon's sister Dani instead, it would take the two of them 7.5 hours to paint the house. How long would it take Dani to complete the job working alone?

5. Refer to Practice 4, during a heavy rain storm, the same leaky barrel takes 2.5 hours to fill. How long would it take the barrel to fill in this rainstorm if it had no leak?

Complex Rational Expressions and Rational Equations
Topic 28.6 Applications of Rational Equations: $D = RT$

Vocabulary
$d = rt$ •   distance   •   rate   •   time   •   formula

1. Distance, rate, and _____ are related by the formula _____.

Step-by-Step Video Notes
Watch the Step-by-Step Video lesson and complete the examples below.

Example	Notes
1 & 2. Solve the formula $d = rt$ for each of the variables below. rate, r $$\frac{d}{\Box} = \frac{rt}{\Box}$$ time, t $$\frac{d}{\Box} = \frac{rt}{\Box}$$	
3. The speed of the current in a river is 5 miles per hour. A boat travels 20 miles downstream (with the current) in this river in the same time that it travels 10 miles upstream (against the current). Find the speed of the boat in still water. Answer:	

Example	Notes
4. At the airport, Mia and Eric are walking at the same speed to catch their flight, but Eric decides to walk on the moving sidewalk, while Mia continues to walk on the stationary sidewalk. If the sidewalk moves at 1 meter per second, and it takes Eric 50 seconds less to walk the 300-meter distance, at what speed are Mia and Eric walking? Answer:	

Helpful Hints
Remember to multiply by the LCD as the first step when solving these types of problems. It can make the calculations much simpler.

Concept Check
1. Dean and Sam are going on a 420 mile road trip in separate cars. It took Sam 1 more hour to complete the trip than it took Dean. Dean averaged 10 miles per hour faster than Sam averaged during the trip. Assume the average speeds are constant. If the speed limit is 65 miles per hour, will both Dean and Sam receive tickets if pulled over? Explain.

Practice
2. Refer to Concept Check 1. How long did the trip take for Dean? For Sam?

4. Refer to Practice 3, what is the speed of the boat if the speed of the current in the river is actually 10 miles per hour?

3. The speed of the current in a river is 8 miles per hour. A boat travels 35 miles downstream (with the current) in this river in the same time that it travels 15 miles upstream (against the current). Find the speed of the boat in still water.

5. At 22 miles per hour, it takes Martha's boat the same amount of time to travel upstream 9 miles as it does to travel downstream 13 miles. What is the speed of the current in the river?

Variation
Topic 29.1 Direct Variation

Vocabulary
direct variation   •   constant of variation   •   function

1. If y varies directly as x, or y is directly proportional to x, then there is a positive constant k such that $y = kx$. In this case, the equation $y = kx$ is called an equation of

    _____.

2. The number k in the direct variation equation $y = kx$ is called the _____ or the constant of proportionality.

Step-by-Step Video Notes
Watch the Step-by-Step Video lesson and complete the examples below.

Example	Notes
1 & 2. Suppose y varies directly as x. Find the constant of variation and the equation of direct variation for the following. $y = 12$ when $x = 3$ $y = -10$ when $x = -2$	

Example	Notes
3 & 4. Find the variation equation in which y varies directly as x and the following are true. Then find y for the given value of x. $y = -6$ when $x = -3$; $x = 10$ $y = 15$ when $x = 20$; $x = 8$	

Helpful Hints

In direct variation, as x increases or decreases, y also increases or decreases, respectively.

Concept Check

1. Suppose y varies directly as x, and $y = 8$ when $x = 2$. Is it possible to find value of x that corresponds to $y = 12$?

Practice

Suppose y varies directly as x. Find the constant of variation and the equation of direct variation for the following for x.

2. $y = 28$ when $x = 7$

Find the variation equation in which y varies directly as x and the following are true. Then find y for the given value of x.

4. $y = -42$ when $x = -6$; $x = 7$

3. $y = -24$ when $x = -8$

5. $y = 9$ when $x = 12$; $x = 8$

Variation
Topic 29.2 Inverse Variation

Vocabulary
inverse variation   •   constant of variation   •   function

1. If y varies inversely as x, or y is inversely proportional to x, then there is a positive constant k such that $y = \dfrac{k}{x}$. In this case, the equation $y = \dfrac{k}{x}$ is called an equation of _____.

2. The number k in the inverse variation equation $y = \dfrac{k}{x}$ is called the _____ or the constant of proportionality.

Step-by-Step Video Notes
Watch the Step-by-Step Video lesson and complete the examples below.

Example	Notes
1 & 2. Find an equation of variation in which y varies inversely as x for each of the following. $y = 5$ when $x = 2$ $y = -4$ when $x = -7$	

Example	Notes
3 & 4. Find an equation of variation in which y varies inversely as x and the following are true. Then find y for the given value of x. $y = -8$ when $x = -7$; $x = 4$ $y = 18$ when $x = \dfrac{1}{2}$; $x = 27$	

Helpful Hints

In inverse variation, as x increases or decreases, y decreases or increases, respectively.

Concept Check

1. Suppose y varies inversely as x, and $y = 3$ when $x = 1$. Can you find x given that $y = 1$?

Practice

Find an equation of variation in which y varies inversely as x for each of the following.

2. $y = 3$ when $x = 12$

Find an equation of variation in which y varies inversely as x and the following are true. Then find y for the given value of x.

4. $y = -12$ when $x = -4$; $x = -6$

3. $y = -11$ when $x = -2$

5. $y = 8$ when $x = 2.5$; $x = -2$

Variation
Topic 29.3 Joint and Combined Variation

Vocabulary
direct variation　•　inverse variation　•　joint variation　•　combined variation

1. When direct variation and inverse variation happen at the same time, we have
   _____.

2. When a variable is proportional to the product of two or more variables, we have
   _____.

Step-by-Step Video Notes
Watch the Step-by-Step Video lesson and complete the examples below.

Example	Notes
1. Suppose y varies jointly with x and z. When $x = 2$ and $z = 9$, then $y = 54$. Find the constant of variation and the equation of joint variation. $y = kxz$ $\Box = k\left(\Box\right)\left(\Box\right)$ $k = \Box$ The equation of joint variation is $y = \Box xz$.	
2. Suppose y varies jointly with x and the square of z. When $x = 3$ and $z = 2$, then $y = 10$. Find y if $x = 5$ and $z = 6$. Answer:	

Example	Notes
4. Suppose y varies directly with the square of x and inversely with z. When $x = 5$ and $z = 4$, then $y = 12\frac{1}{2}$. Find y if $x = 1$ and $z = 7$.	

Answer:

Helpful Hints

If y varies directly as x and z, then there is a positive constant k such that $y = kxz$. In this case, the equation is called an equation of joint variation.

If y varies directly as x and inversely as z, then there is a positive constant k such that $y = \dfrac{kx}{z}$. In this case, the equation is called an equation of combined variation.

Concept Check

1. Suppose y varies directly with x and inversely with z. If x and z are both doubled, what happens to y? Explain.

Practice

Suppose y varies jointly with the cube of x and the square root of z.

2. When $x = 3$ and $z = 16$, then $y = 27$. Find the constant of variation and the equation of joint variation.

Suppose y varies jointly with x and z, and inversely as p.

4. When $x = 4$, $z = 3$, and $p = 2$, then $y = 42$. Find the constant of variation and the equation of combined variation.

3. Refer to Practice 2, find y if $x = 2$ and $z = 4$.

5. Refer to Practice 4, find y if $x = 8$, $z = 5$, and $p = 14$.

Variation
Topic 29.4 Applications of Variation

Vocabulary
direct variation   •   inverse variation   •   joint variation   •   combined variation

1. The equation $y = \dfrac{k}{x}$, where k is a positive constant, is called an equation of

    _____.

Step-by-Step Video Notes
Watch the Step-by-Step Video lesson and complete the examples below.

Example	Notes
1. The distance d that a spring stretches varies directly with the weight w of the object hung on the spring. If a 10-pound fish stretches the spring in a fishing scale 6 cm, how far will a 35-pound fish stretch this spring? Answer:	
3. If the voltage in an electric circuit is kept at the same level, the current I varies inversely with the resistence R. The current measures 40 amperes when the resistance is 300 ohms. Find the current when the resistance is 100 ohms. Answer:	

Example	Notes
5. The load capacity U of a cylindrical concrete column varies directly with the diameter d of the column raised to the fourth power, and inversely with the square of the length l of the column. If the diameter of a column is 1 foot, and the length is 10 feet, the load capacity is 184 pounds. Find the load capacity if the diameter is 1.5 feet and the length is 12 feet. Answer:	

Helpful Hints

If y varies jointly as x and z, then there is a positive constant k such that $y = kxz$. In this case, the equation is called an equation of joint variation.

If y varies directly as x and inversely as z, then there is a positive constant k such that $y = \dfrac{kx}{z}$. In this case, the equation is called an equation of combined variation.

Concept Check

1. Refer to Example 5, if the length of the column were increased, would it increase or decrease the load capacity of the column? Explain.

Practice

2. The revenue that Harry will earn with his lemonade stand varies directly with the number of cups he sells. Suppose he earns $40.25 by selling 35 cups. How much would he earn if he sold 47 cups?

3. The number of refrigerators sold at a department store in a week varies inversely with the price. If 10 are sold at $400 per unit, how many would sell if the price were raised to $500?

4. The volume of a pyramid with a square base varies jointly as the height and the square of the side length of the base. The volume of a pyramid with a height of 9 m and a base side length of 2 m is 12 m^3. Find the volume of a pyramid with a height of 5 m and a base side length of 3 m. (The units for volume are m^3.)

Name: _____   Date: _____

Instructor: _____   Section: _____

Roots and Radicals
Topic 30.1 Square Roots

Vocabulary
square root   •   principal square root   •   negative square root

1. The _____ is one of two identical factors of a number.

2. The radical symbol $\sqrt{}$ is used to denote the _____ of a number.

Step-by-Step Video Notes
Watch the Step-by-Step Video lesson and complete the examples below.

Example	Notes
1–4. Find the square roots.	
$\sqrt{100}$	
$\sqrt{\dfrac{9}{25}}$	
$-\sqrt{49}$	
$\sqrt{0}$	
7 & 8. Find the square roots.	
$\sqrt{49x^2}$	
$\sqrt{(-6x)^2}$	

Example	Notes
9 & 10. Find the square roots. $\sqrt{-144}$ $-\sqrt{-9}$	

12. Approximate $\sqrt{75}$ by finding the two consecutive whole numbers that the square root lies between. Answer:	

Helpful Hints
If a variable appears in the radicand, assume it represents positive numbers only.

The square root of a negative number is not a real number.

Not every positive number has a rational square root. You can use a calculator to approximate the square roots of such numbers.

Concept Check
1. The value of $\sqrt{22}$ is between what two whole numbers?

Practice
Find the square roots.

2. $\sqrt{144}$

4. $\sqrt{64z^2}$

3. $-\sqrt{\dfrac{36}{121}}$

5. $\sqrt{-\dfrac{1}{4}}$

Roots and Radicals
Topic 30.2 Higher-Order Roots

Vocabulary

square root   •   cube root   • n^{th} root

1. The _____ of a number is one of three identical factors.

2. The _____ is one of n identical factors of a number.

Step-by-Step Video Notes
Watch the Step-by-Step Video lesson and complete the examples below.

Example	Notes
1–3. Find the cube roots. $\sqrt[3]{27x^3}$ $-\sqrt[3]{\dfrac{8}{27}}$ $\sqrt[3]{-1}$	
4 & 5. Find the indicated roots. $\sqrt[5]{32}$ $\sqrt[4]{-81}$	

Example	Notes
6–8. Simplify the radicals. $\sqrt[5]{x^{10}}$ $\sqrt[4]{y^{12}}$ $\sqrt[4]{a^{24}b^{28}}$	
9 & 10. Simplify the radicals. $\sqrt[3]{-64x^6}$ $\sqrt[4]{10,000x^4}$	

Helpful Hints

A higher-order root is found using the radical sign $\sqrt[n]{a}$ where n is the index of the radical and a is called the radicand. To find an nth root, we can divide the exponent of the radicand by the index.

The cube root of a negative number is a negative number. If the index is even, the radicand must be nonnegative for the root to be a real number.

Concept Check

1. Find three different sets of whole number values for n and a if $\sqrt[n]{a} = 2$ and $n \geq 3$.

Practice

Simplify the radicals.

2. $\sqrt[3]{-27}$

3. $\sqrt[6]{x^{30}}$

4. Fill in the table below.

x	x^2	x^3	x^4	x^5
1	1	1	1	1
2	4		16	
3				243
4			256	
5		125		3125

Roots and Radicals
Topic 30.3 Simplifying Radical Expressions

Vocabulary

square root   •   product rule for radicals   •   prime factorization

1. The _____ states that for all nonnegative real numbers a, b, and n,
   $\sqrt[n]{a} \cdot \sqrt[n]{b} = \sqrt[n]{ab}$ and $\sqrt[n]{ab} = \sqrt[n]{a} \cdot \sqrt[n]{b}$.

Step-by-Step Video Notes
Watch the Step-by-Step Video lesson and complete the examples below.

Example	Notes
1. Simplify the radical $\sqrt{50}$. Factor the radicand. If possible write the radicand as a product of a perfect square. Use the Product Rule to separate the factors. Answer:	
2 & 3. Simplify the radicals. $\sqrt{8}$ $\sqrt{48}$	

Example	Notes
5 & 6. Simplify the radicals. Assume all variables represent nonnegative values. If the answer is not a real number, say so. $\sqrt{x^5}$ $\sqrt[3]{y^{17}}$	
7 & 8. Simplify the radicals. Assume all variables represent nonnegative values. If the answer is not a real number, say so. $\sqrt{24x^3}$ $\sqrt[3]{-16x^{11}}$	

Helpful Hints
When simplifying, assume that all variables inside radicals represent nonnegative values.

When simplifying radicals, you can find the perfect root factors in a more difficult problem by looking at the prime factorizations of the radicands.

Concept Check
1. Of $\sqrt[3]{a^2}$, $\sqrt[3]{a^4}$, and $\sqrt[3]{-a^9}$, which expression cannot be simplified?

Practice
Simplify. Assume all variables represent nonnegative values.

2. $\sqrt{75}$ 4. $\sqrt[3]{-24}$

3. $\sqrt{m^7}$ 5. $\sqrt[3]{-250y^4}$

Roots and Radicals
Topic 30.4 Rational Exponents

Vocabulary
rational exponents   •   product rule for exponents   •   radical notation

1. _____ can be written as radicals.

2. The _____ states that $x^a \cdot x^b = x^{a+b}$.

Step-by-Step Video Notes
Watch the Step-by-Step Video lesson and complete the examples below.

Example	Notes
1 & 2. Write in radical notation. Simplify, if possible. $9^{1/2}$ $x^{1/3}$	
3 & 4. Write in radical notation. Simplify, if possible. $6^{2/3}$ $x^{3/2}$	

Example	Notes
5 & 6. Use radical notation to rewrite each expression. Simplify if possible. $\left(\dfrac{1}{9}\right)^{3/2}$ $(3x)^{1/3}$	
9. Use the rules of exponents to simplify. $\dfrac{6^{1/3}}{6^{4/3}}$	

Helpful Hints

If m and n are integers greater than 1, with $\dfrac{m}{n}$ in simplest form, then $a^{m/n} = \sqrt[n]{a^m} = \left(\sqrt[n]{a}\right)^m$, as long as $\sqrt[n]{a}$ is a real number.

If an exponential expression has a negative rational exponent, write the expression with a positive exponent before evaluating the rational exponent.

Concept Check

1. Express $\sqrt[5]{32x^{10}}$ with rational exponents and simplify if possible.

Practice

Write in radical form. Simplify if possible.

2. $(4x)^{2/3}$

3. $8^{5/3}$

Simplify.

4. $x^{2/3} \cdot x^{1/4}$

5. $16^{-3/2}$

Roots and Radicals
Topic 30.5 More on the Pythagorean Theorem

Vocabulary

hypotenuse   •   leg   •   the Pythagorean Theorem

1. The side opposite the right angle in a right triangle is called the _____.

2. In a right triangle, _____ states that the sum of the squares of the legs is equal to the square of the hypotenuse, or $leg^2 + leg^2 = hypotenuse^2$.

Step-by-Step Video Notes
Watch the Step-by-Step Video lesson and complete the examples below.

Example	**Notes**
1. Find the length of the hypotenuse. *[right triangle with hypotenuse c, one leg 3 feet, other leg 4 feet]* Answer:	
2. Find the length of the missing leg. *[right triangle with leg a, hypotenuse 17 m, other leg 15 m]* Answer:	
3. Find the length of the hypotenuse. *[right triangle with leg 9 inches, leg 6 inches, hypotenuse c]* Answer:	

Example	Notes
4 A slanted roof rises 5 feet vertically from the edge of the roof to the top. The roof covers 12 horizontal feet. How long is the slanted surface of the roof? 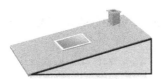	

Answer:

Helpful Hints

The Pythagorean Theorem is often presented as $a^2 + b^2 = c^2$, where a and b represent the legs of a right triangle and c represents the hypotenuse.

When using the Pythagorean Theorem, if the missing length is not a rational number, then express it either as a simplified radical or use a calculator to approximate it to a certain number of places. A given problem will usually specify whether an approximated or exact answer is required.

Concept Check

1. Why can't you always use the Pythagorean Theorem to find the missing side of a triangle with a side of length 2 m and a side of length 8 m?

Practice

Find the missing length in each right triangle.

2.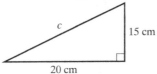

3.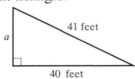

4.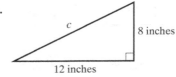

Roots and Radicals
Topic 30.6 The Distance Formula

Vocabulary
the Distance Formula   •   the Pythagorean Theorem

1. To calculate the distance between any two points (x_1, y_1) and (x_2, y_2) on a graph, use

    _____, which states $d = \sqrt{(x_2 - x_1)^2 + (y_2 - y_1)^2}$.

Step-by-Step Video Notes
Watch the Step-by-Step Video lesson and complete the examples below.

Example	Notes
1. Find the length of the line segment between the two points shown. 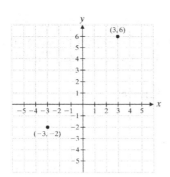 Answer:	
3. Find the distance between $(-3, 2)$ and $(1, 5)$. 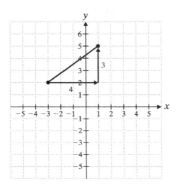 Answer:	

Example	Notes
4 & 5. Use the Distance Formula to find the distance between the given points. $(6, -3)$ and $(1, 9)$ $(2, 1)$ and $(6, 8)$	
6. Use the Distance Formula to find the distance between the points $(-13, -8)$ and $(-3, -3)$. Answer:	

Helpful Hints

Notice in example 3 that the horizontal and vertical distances between the two points make up the legs of a right triangle. The hypotenuse of this right triangle is the distance between the two points. This is how the Distance Formula is derived from the Pythagorean Theorem.

Be careful to avoid sign errors when finding the differences between the x- and y-coordinates of the points.

Concept Check
1. Will the Distance Formula also work for two points on the same horizontal line or the same vertical line? Explain.

Practice
Find the distance between the two points shown on the graph.

2.

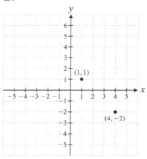

Use the Distance Formula to find the distance between the given points.

3. $(-3, -4)$ and $(5, 11)$

4. $(4, 7)$ and $(8, 13)$

Name: _____   Date: _____

Instructor: _____   Section: _____

Operations of Radical Expressions
Topic 31.1 Introduction to Radical Functions

Vocabulary
radical function   •   domain   •   square root   •   evaluate

1. To find the _____ of a function, find the values of x that make $f(x)$ a real number.

Step-by-Step Video Notes
Watch the Step-by-Step Video lesson and complete the examples below.

Example	Notes
1–3. If $f(x) = \sqrt{2x+4}$, find the function values. $f(2.5)$ $f(-3)$ $f(-2)$	
4. Find the domain of $f(x) = \sqrt{x+2}$. Solve $x+2 \geq 0$ to find the domain. Answer:	
5. Graph $f(x) = \sqrt{x+2}$. 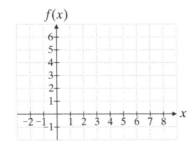	

Example	Notes
7. Graph $f(x) = \sqrt[3]{x}$. 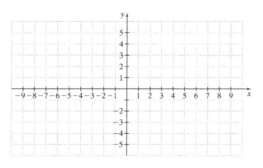	

Helpful Hints

To evaluate a function at a certain value, substitute that value for the variable in the expression and simplify.

Remember that the square root of a negative number is not a real number, while the cube root of a negative number is a negative number.

Concept Check

1. Compare the domains of $f(x) = \sqrt{x-1}$, $g(x) = \sqrt{x-2}$, and $h(x) = \sqrt{x-3}$. What do you think the domain of $p(x) = \sqrt{x-a}$, where a is any real number would be?

Practice

2. If $f(x) = \sqrt{3x-5}$, find the function values. State the domain of f.

$$f\left(-\frac{5}{3}\right)$$

$$f\left(\frac{5}{3}\right)$$

3. If $f(x) = \sqrt{x+10}$, find the function values. State the domain of f.

$$f(-6)$$

$$f(6)$$

4. Graph $f(x) = \sqrt[3]{x-1}$. State the domain.

Operations of Radical Expressions
Topic 31.2 Adding and Subtracting Radical Expressions

Vocabulary
like radicals   •   like terms   •   radical expressions

1. Radicals with the same radicand and the same index are called _____.

Step-by-Step Video Notes
Watch the Step-by-Step Video lesson and complete the examples below.

Example	Notes
1–3. Combine like terms. $7x + 9x$ $13xy^2 + 11x^2y$ $5x^{1/2} + 2x^{1/2}$	
4–6. Simplify, if possible. Assume that all variables are nonnegative real numbers. $5\sqrt{3} + 7\sqrt{3}$ $3\sqrt{2} - 9\sqrt{2}$ $-6\sqrt{xy} + 2\sqrt[3]{xy}$	

Example	Notes
8. Simplify. $5\sqrt{3} - \sqrt{27} + 2\sqrt{32}$	
10. Simplify. Assume that all variables are nonnegative real numbers. $3x\sqrt[3]{54x^4} - 3\sqrt[3]{16x^7}$	

Helpful Hints

A radical is written using the radical sign $\sqrt[n]{a}$ where n is the index of the radical and a is called the radicand.

For all nonnegative real numbers a, b, and n, where n is an integer greater than 1, $\sqrt[n]{ab} = \sqrt[n]{a} \cdot \sqrt[n]{b}$.

Concept Check

1. Are $x\sqrt{7}$ and $7\sqrt{x}$ like radicals? Explain why or why not.

Practice

Simplify. Assume variables represent nonnegative numbers.

2. $11\sqrt{7} - 12\sqrt{7}$

4. $-\sqrt{12} + 6\sqrt{27} - 4\sqrt{28}$

3. $\sqrt{48} + \sqrt{75}$

5. $\sqrt{72x} - 2x\sqrt{3} - 5\sqrt{2x} + x\sqrt{27}$

Operations of Radical Expressions
Topic 31.3 Multiplying Radical Expressions

Vocabulary
Distributive Property   •   Product Rule for Radicals   •   FOIL method

1. The _____ states that for all nonnegative real numbers a, b,

    and n, where n is an integer greater than 1, $\sqrt[n]{a} \cdot \sqrt[n]{b} = \sqrt[n]{ab}$.

Step-by-Step Video Notes
Watch the Step-by-Step Video lesson and complete the examples below.

Example	Notes
2. Multiply. $\sqrt{5} \cdot \sqrt{3}$ Use the Product Rule for Radicals. Answer:	
4. Multiply. $\left(\sqrt{12}\right)\left(-5\sqrt{3}\right)$ Use the Product Rule for Radicals. Simplify, if possible. Answer:	

Example	Notes
6. Multiply. $\left(\sqrt{2}+3\sqrt{5}\right)\left(2\sqrt{2}-\sqrt{5}\right)$ Answer:	
8. Simplify. Assume all variables represent nonnegative numbers. $\left(\sqrt{7}+\sqrt{3x}\right)^2$ Answer:	

Helpful Hints

Apply multiplicative properties like the Distributive Property, the Product Rules, the FOIL method, the difference of two squares, and squaring binomials when multiplying radical expressions, just as you would with other types of expressions.

Recall that $\left(\sqrt{x}\right)^2 = x$.

Concept Check

1. Addison claims that $\left(\sqrt{-3}\right)\left(\sqrt{-3}\right)=\left(\sqrt{9}\right)=3$. Is she correct? Explain why or why not.

Practice

Multiply. Assume all variables represent nonnegative numbers.

2. $\left(4x\sqrt{2y}\right)\left(7x\sqrt{8}\right)$

4. $\left(4\sqrt{5}+3\sqrt{2}\right)\left(4\sqrt{5}-3\sqrt{2}\right)$

3. $\sqrt{3x}\left(2\sqrt{6x}+7\sqrt{12}\right)$

5. $\sqrt[3]{7x}\left(\sqrt[3]{49x^2}-5\sqrt[3]{2x}\right)$

Operations of Radical Expressions
Topic 31.4 Dividing Radical Expressions

Vocabulary
Quotient Rule for Radicals   •   radical expressions

1. The _____ states that for all nonnegative real numbers a, b, and n, where n is an integer greater than 1, and $b \neq 0$, $\dfrac{\sqrt[n]{a}}{\sqrt[n]{b}} = \sqrt[n]{\dfrac{a}{b}}$.

Step-by-Step Video Notes
Watch the Step-by-Step Video lesson and complete the examples below.

Example	Notes
1. Divide. $$\dfrac{\sqrt{75}}{\sqrt{3}}$$ Answer:	
2 & 3. Divide. $$\dfrac{\sqrt{9}}{\sqrt{16}}$$ $$\dfrac{\sqrt{72}}{\sqrt{8}}$$	

Example	Notes
4 & 5. Divide. $\sqrt[3]{\dfrac{-a^9}{b^6}}$ $\dfrac{\sqrt[3]{x^8}}{\sqrt[3]{x^5}}$	

7. Divide. Assume variables represent nonnegative values.

$$\frac{\sqrt{25x^5y^2}}{\sqrt{144x^3y^4}}$$

Answer:

Helpful Hints
The Quotient Rule for Radicals can be very flexible. You can simplify the numerators and/or denominators first, or you can divide the radicands first, depending on the situation.

Concept Check
1. Simplify Example 5 in a different way than you did originally. Do you get the same answer?

Practice
Divide. Assume all variables represent nonnegative numbers.

2. $\dfrac{\sqrt{150}}{\sqrt{6}}$

3. $\dfrac{\sqrt{63a^9b^7}}{\sqrt{7a^3b}}$

4. $\dfrac{\sqrt{25a^6}}{\sqrt{81b^{12}c^4}}$

5. $\sqrt{\dfrac{147a^2b^6}{3a^8b^4}}$

Operations of Radical Expressions
Topic 31.5 Rationalizing the Denominator

Vocabulary
rationalizing the denominator • Identity Property of Multiplication • conjugates

1. Performing operations to remove a radical from a denominator is called

_____.

Step-by-Step Video Notes
Watch the Step-by-Step Video lesson and complete the examples below.

Example	**Notes**
1. Simplify by rationalizing the denominator. $\dfrac{1}{\sqrt{2}}$ Answer:	
3. Simplify by rationalizing the denominator. $\sqrt[3]{\dfrac{2}{3x^2}}$ Answer:	
4. Simplify by rationalizing the denominator. $\dfrac{5}{3+\sqrt{2}}$ Answer:	

Example	Notes
5. Simplify by rationalizing the denominator. $$\dfrac{\sqrt{7}+\sqrt{3}}{\sqrt{7}-\sqrt{3}}$$ Answer:	

Helpful Hints

A rational expression is not considered to be in simplest form if there is an irrational expression in the denominator.

When rationalizing a denominator, you can simplify the radical in the denominator first, and then rationalize the denominator, or you can rationalize first, and then simplify the fraction.

The product of two conjugates is always rational.

Concept Check

1. Explain why $\dfrac{2}{1-\sqrt{2}}$ cannot be rationalized by multiplying by $\dfrac{\sqrt{2}}{\sqrt{2}}$.

Practice

Simplify by rationalizing the denominator. Assume variables represent nonnegative numbers.

2. $\dfrac{2x}{\sqrt{7}}$

4. $\sqrt[3]{\dfrac{4}{3x^2}}$

3. $\dfrac{5}{\sqrt{75}}$

5. $\dfrac{8}{3-\sqrt{5}}$

Operations of Radical Expressions
Topic 31.6 Solving Radical Equations

Vocabulary
Squaring Property of Equality   •   reverse operations

1. For all real numbers a and b, if $a = b$, then $a^2 = b^2$ because of the _____.

Step-by-Step Video Notes
Watch the Step-by-Step Video lesson and complete the examples below.

Example	Notes
1. Solve. $\sqrt{x} = 7$ Square both sides of the equation. Check. Answer:	
3. Solve. $3 + \sqrt{x-5} = 15$ Answer:	

Example	Notes
5. Solve. $-6+\sqrt[3]{x}=-10$ Answer:	
6. When a car traveling on wet pavement at a speed V in miles per hour stops suddenly, it will produce skid marks of length S feet according to the formula $V=2\sqrt{3S}$. Solve the equation for S, and then use this result to find the length of the skid mark if the car is traveling at 30 miles per hour. Answer:	

Helpful Hints

The first step in solving a radical equation is to get a radical on one side of the equation by itself.

Always check all possible solutions to a radical equation to make sure they work in the original equation. There may be extraneous roots that do not work in the original equation.

Concept Check

1. If k is a positive constant, does the equation $\sqrt{x}=-k$ have any solutions? Explain.

Practice

Solve.

2. $\sqrt{x}=-3$

3. $3-\sqrt{x+7}=-9$

4. $x=\sqrt{5x+14}$

5. $\sqrt[3]{x-8}=3$

Solving Quadratic Equations
Topic 32.1 Introduction to Solving Quadratic Equations

Vocabulary
quadratic equation   •   solution   •   parabola   • x-intercept

1. A(n) _____ is an equation of the form $ax^2 + bx + c = 0$ where $a \neq 0$.

2. A(n) _____ of the graph of an equation is a point where the curve crosses the x-axis, or where the value of y is 0.

Step-by-Step Video Notes
Watch the Step-by-Step Video lesson and complete the examples below.

Example	Notes
1–3. Write each quadratic equation in standard form. $7x^2 = 5x + 8$ $6x - 2x^2 = -3$ $4x^2 = 9$	
4–6. Determine the number of solutions for each quadratic equation graphed below. 	

Example	**Notes**
7 & 8. Find the real solution(s) of each quadratic equation, if any exist, by using the graph of the equation. 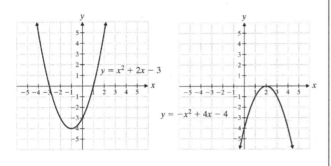	

Helpful Hints

The graph of a quadratic equation $y = ax^2 + bx + c$ is called a parabola.

A quadratic equation may have two real solutions, one real solution, or no real solutions.

Concept Check

1. How can you use the graph of a quadratic equation to determine the real solutions of the equation?

Practice

Write each quadratic equation in standard form.

2. $x^2 + 9 = -6x$

3. $2x + 3 = 5x^2$

Find the real solution(s) of each quadratic equation by using the graph of the equation.

4.

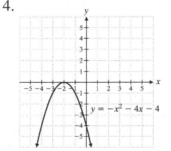

5.

Name: _____   Date: _____
Instructor: _____   Section: _____

Solving Quadratic Equations
Topic 32.2 Solving Quadratic Equations by Factoring

Vocabulary
standard form   •   factor   •   quadratic expression

1. To solve a quadratic equation by factoring, first make sure the equation is in the
   _____, $ax^2 + bx + c = 0$.

Step-by-Step Video Notes
Watch the Step-by-Step Video lesson and complete the examples below.

Example	Notes
1. Solve $x^2 - 9x = 0$ by factoring. Factor completely. Set each factor equal to zero. Solve each equation. Check each solution. Answer:	
2. Solve $5x^2 - 14x - 3 = 0$ by factoring. Answer:	

Example	Notes
3. Solve $9x^2 = 24x - 15$ by factoring.	
Answer:	
4. Solve $x^2 - 2x = -1$ by factoring.	
Answer:	

Helpful Hints
When solving a quadratic equation by factoring, set each factor containing a variable equal to zero. Remember, if you factor out the variable or a monomial term containing the variable as the GCF, then 0 is one of the solutions of the equation.

Concept Check
1. Rewrite $y = (x - 7)^2$ by setting each factor equal to zero to solve. How many solutions does this equation have?

Practice
Solve by factoring.

2. $3x^2 + 15x = 0$

3. $5x^2 + 3x - 14 = 0$

4. $7x^2 + 84 = 56x$

5. $x^2 + 10x + 25 = 0$

Solving Quadratic Equations
Topic 32.3 Solving Quadratic Equations Using the Square Root Property

Vocabulary
Square Root Property   •   quadratic equation

1. A _____ is an equation of the form $ax^2 + bx + c = 0$ where $a \neq 0$.

Step-by-Step Video Notes
Watch the Step-by-Step Video lesson and complete the examples below.

Example	Notes
2. Solve $x^2 = 48$ for x. Use the Square Root Property. Simplify and check your solutions. Answer:	
3. Solve $x^2 = -4$ for x. Answer:	

Example	Notes
4. Solve $3x^2 + 2 = 77$ for x.	
Answer:	
5. Solve $(4x - 1)^2 = 5$ for x.	
Answer:	

Helpful Hints

The notation $\pm\sqrt{a}$ is a shorthand way of writing "$+\sqrt{a}$ or $-\sqrt{a}$".

The Square Root Property can only be used to solve an equation of the form $x^2 = a$ if $a \geq 0$. If $a < 0$, the equation has no real solutions.

Concept Check

1. Mark uses the Square Root Property to solve the equation $x^2 + 4 = 0$ and incorrectly gets $x = \pm 2$ as his answer. Explain his error and state the correct answer.

Practice

Solve for x.

2. $x^2 = 28$

4. $6x^2 + 2 = 98$

3. $x^2 = 196$

5. $(2x - 3)^2 = 25$

Solving Quadratic Equations
Topic 32.4 Solving Quadratic Equations by Completing the Square

Vocabulary
completing the square • the Square Root Property • perfect square trinomial

1. To solve an equation like $x^2 + bx = c$, we can add a constant to both sides of the equation so that the left side becomes a perfect square trinomial. This method is called

   _____.

Step-by-Step Video Notes
Watch the Step-by-Step Video lesson and complete the examples below.

Example	Notes
1 & 2. Fill in the blanks to create a perfect square trinomial. $x^2 + 8x + \underline{} = \left(x + \underline{}\right)^2$ $x^2 - 12x + \underline{} = \left(x - \underline{}\right)^2$	
3. Solve $x^2 + 2x = 3$ by filling in the blanks and then using the Square Root Property. $x^2 + 2x + \underline{} = 3 + \underline{}$ Answer:	

Example	Notes
4. Solve $x^2 + 4x - 5 = 0$ by completing the square. Answer:	
5. Solve $x^2 + 6x + 1 = 0$ by completing the square. Answer:	

Helpful Hints

Recall that when a binomial with is squared using FOIL, the coefficient of x is twice the product of coefficient of x in the binomial and the constant of the binomial.

If the equation you are solving contains fractions or if the coefficient of x is odd, then the equation will be more easily solved by using a method other than completing the square.

Concept Check

1. Which method would you use to solve $x^2 - 8x = -9$? Which method would you use to solve $x^2 - 8x = -16$?

Practice

Solve by completing the square.

2. $x^2 + 5x - 6 = 0$

4. $2x^2 + 16x - 96 = 0$

3. $x^2 + 12x + 4 = 0$

5. $x^2 - 4x - 16 = 0$

Name: _____   Date: _____

Instructor: _____   Section: _____

Solving Quadratic Equations
Topic 32.5 Solving Quadratic Equations Using the Quadratic Formula

Vocabulary
standard form   •   Square Root Property   •   quadratic formula

1. For all quadratic equations in the form $ax^2 + bx + c = 0$, you can solve by using the

   _____, which states that $x = \dfrac{-b \pm \sqrt{b^2 - 4ac}}{2a}$.

Step-by-Step Video Notes
Watch the Step-by-Step Video lesson and complete the examples below.

Example	Notes
1. Solve $3x^2 - x - 2 = 0$ by using the quadratic formula. Identify a, b, and c. Substitute a, b, and c into the quadratic formula. Simplify. Answer:	
2. Solve $x^2 = 6x$ by using the quadratic formula. Answer:	

Example	Notes
3. Solve $4x^2 + 25 = 20x$ by using the quadratic formula. Answer:	
5. Solve $x^2 + 4x - 8 = 0$ by using the quadratic formula. Answer:	

Helpful Hints

The quadratic formula is the only method of solving that works for every quadratic equation.

Be sure the equation is in standard form before you identify a, b, and c. If b is positive, then $-b$ in the formula is negative, but if b is negative, then $-b$ is positive in the formula.

Sometimes the value of b or c will be equal to 0. If there is no x term, then the value of b is equal to 0; if there is no constant, then the value of c is equal to 0.

Concept Check

1. How many real solutions does a quadratic equation in the form $ax^2 + bx + c = 0$ have if $b^2 - 4ac = 0$? If $b^2 - 4ac < 0$?

Practice

Solve by using the quadratic formula.

2. $x^2 - 8x + 7 = 0$

4. $9x^2 = 6x - 1$

3. $x^2 + 8x = -2 + 3x$

5. $6x = 4x^2 + 3$

Solving Quadratic Equations
Topic 32.6 Applications with Quadratic Equations

Vocabulary

$S(t) = -5t^2 + vt + h$ •   quadratic formula   •   right triangle

1. In a _____, the sum of the squares of its legs is equal to the square of the hypotenuse, or $a^2 + b^2 = c^2$.

Step-by-Step Video Notes

Watch the Step-by-Step Video lesson and complete the examples below.

Example	Notes
1. A tennis ball is thrown upward with an initial velocity of 10 meters per second. Suppose that the initial height above the ground is 4 meters. $S(t) = -5t^2 + vt + h$ $\quad = -5t^2 + \boxed{} \cdot t + \boxed{}$ Find the height, S, of the ball after 1 second. $S(1) = -5\left(\boxed{}\right)^2 + 10\left(\boxed{}\right) + 4$ The height of the ball after 1 second is $\boxed{}$ meters. At what time, t, will the ball hit the ground? Round your answer to the nearest hundredth. Remember, the function for when an object is thrown or launched is given by $S(t) = -5t^2 + vt + h$. $S(t) = 0$, find t. Use the quadratic formula.	

Example	Notes

2. A rocket is fired upward with an initial velocity, v, of 110 meters per second. The quadratic function $S(t) = -5t^2 + 110t$ can be used to find the heights, S, of the rocket, in meters, at any time, t, in seconds.

Find the height of the rocket 8 seconds after it takes off.

Find how many second it takes for the rocket to reach 500 meters. Round your answers to the nearest hundredth.

$S(t) = 500$, find t. Use the quadratic formula.

3. Find the lengths of the sides of the triangle. Round your answers to the nearest tenth.

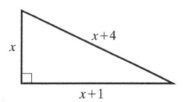

Use the Pythagorean Theorem to find x.

Use the value of x to get the lengths of the three sides of the triangle.

Answer:

Example	Notes

4. A backyard pool has a concrete walkway around it that is 4 feet wide on all sides. The total area of the pool and the walkway is 1015 ft². If the length of the pool is 6 feet longer than the width, find the dimensions of the pool.

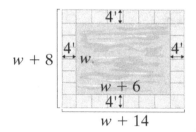

Answer:

Helpful Hints

When an object is thrown or launched upward its approximate height in meters is given by the quadratic function $S(t) = -5t^2 + vt + h$, where $S(t)$ is the height in meters, v is the initial velocity in meters per second, t represents the time in seconds after the object is thrown or launched, and h represents the initial height of the object.

For all quadratic equations in the form $ax^2 + bx + c = 0$, $x = \dfrac{-b \pm \sqrt{b^2 - 4ac}}{2a}$.

In application problems that involve quadratic equations, be sure to carefully consider the solutions to the quadratic equation. Often, negative solutions will not make sense in the context of the problem and should be ignored.

Concept Check

1. Consider a rectangle with width $x + 6$ units, length $x + 7$ units, and an area of 20 square units. Using the formula for the area of a rectangle, the equation $(x+6)(x+7) = 20$ is obtained. Explain why a negative value of x is an acceptable solution to the equation in this case.

Practice

2. A rocket is fired upward with an initial velocity, v, of 90 meters per second. The quadratic function $S(t) = -5t^2 + 90t$ can be used to find the heights, S, of the rocket in meters at any time, t, in seconds. Find the height of the rocket 5 seconds after it takes off.

4. Find the lengths of the sides of the triangle. Round your answers to the nearest tenth.

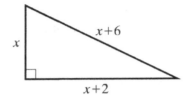

3. Refer to Practice 2, find how many seconds it takes for a rocket to reach a height of 337 meters. Round your answers to the nearest hundredth.

5. A backyard garden has an ornate brick border around it that is 6 inches wide on all sides. The total area of the garden and the border is 3888 square inches. If the length of the garden is 72 inches longer than the width, find the dimensions of the garden.

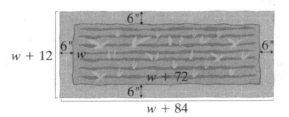

Complex Numbers and More Quadratic Equations
Topic 33.1 Complex Numbers

Vocabulary
imaginary numbers   •   complex numbers   •   powers of i

1. The set of _____ consists of numbers of the form bi where b is a real number, and $b \neq 0$.

2. _____ are numbers of the form $a + bi$, where a and b are real numbers.

Step-by-Step Video Notes
Watch the Step-by-Step Video lesson and complete the examples below.

Example	Notes
1 & 2. Simplify. $\sqrt{-49}$ $\sqrt{-8}$	
3 & 4. Evaluate the powers of i. i^9 i^{15}	
5. Add or subtract. $$(5 + 6i) + (6 - 3i) = \square + \square + \square - \square$$ $$= (\square + \square) + (\square - \square)$$ $$= \square + \square$$ Answer:	

Example	Notes
8. Multiply. $(7-6i)(2+3i)$ Answer:	
10. Divide. $\dfrac{7+i}{3-2i}$ Answer:	

Helpful Hints

For all positive real numbers a, $\sqrt{-a} = \sqrt{-1} \cdot \sqrt{a} = i\sqrt{a}$.

Note that $i^{4n} = 1$ where $n \neq 0$. Keep this in mind when evaluating powers of i.

Concept Check

1. Why is $1 + \pi \cdot i^4$ a real number while $\pi + i^5$ is not? Are they complex numbers? If so, what are their real and imaginary parts?

Practice

Add, subtract, multiply, or divide as indicated.

2. $(1.28 + 2i) + (11 - 5i)$

4. $(6 - 2i)(2 + 7i)$

3. $\left(\dfrac{3}{2} + 8i\right) - (3 - 6i)$

5. $\dfrac{1 + 4i}{7 - 3i}$

Complex Numbers and More Quadratic Equations
Topic 33.2 The Discriminant in the Quadratic Formula

Vocabulary
rational numbers   •   irrational numbers   •   imaginary number
complex numbers   •   discriminant

1. The _____ of a quadratic equation is the simplified value of $b^2 - 4ac$.

2. _____ are numbers of the form $a + bi$ where a and b are real numbers.

Step-by-Step Video Notes
Watch the Step-by-Step Video lesson and complete the examples below.

Example	Notes
1. Use the discriminant to determine how many and what kind of solutions the quadratic equation will have. Do not solve the equation. $5x^2 - 8x = 4$ $5x^2 \qquad -8x \qquad -\square \ = \square$ $\uparrow \qquad\quad \uparrow \qquad\quad \uparrow$ $a = \square \quad b = \square \quad c = \square$ Compute the discriminant. $b^2 - 4ac$ $(\square)^2 - 4(\square)(\square)$ Identify how many and what kind of solutions the quadratic equations will have. $\square$ is a positive, perfect square. Answer:	

Example	Notes
3 & 4. Use the discriminant to determine how many and what kind of solutions the quadratic equations will have. Do not solve the equations. $4x^2 + 81 = 0$ $3x^2 + 4x - 2 = 0$	

Helpful Hints

For all quadratic equations in the form $ax^2 + bx + c = 0$, $x = \dfrac{-b \pm \sqrt{b^2 - 4ac}}{2a}$.

If the discriminant is positive and a perfect square, then there are two real, rational solutions to the quadratic equation. If it is positive, but not a perfect square, then there are two real, irrational solutions. If it is zero, then there is one real, rational solution. If it is negative, then there are no real solutions and two non-real complex solutions.

Concept Check

1. Consider the equation $x^2 + bx + 81 = 0$. For what value(s) of b is there one real, rational solution? Explain.

Practice

Use the discriminant to determine how many and what kind of solutions the quadratic equations will have. Do not solve the equations.

2. $x^2 - 6x + 9 = 0$ 4. $5x^2 + 16 = 9x$

3. $7x - 2 = -4x^2$ 5. $x^2 - 3 = 0$

Complex Numbers and More Quadratic Equations
Topic 33.3 Solving Quadratic Equations with Real or Complex Number Solutions

Vocabulary
quadratic formula • discriminant • imaginary number • complex numbers

1. The _____, i, is defined as $i = \sqrt{-1}$, and $i^2 = -1$.

2. The _____ tells the nature of the solutions to a quadratic equation.

Step-by-Step Video Notes
Watch the Step-by-Step Video lesson and complete the examples below.

Example	Notes
1. Solve $x^2 - 8x + 10 = 0$ using the quadratic formula. Identify a, b, and c. Substitute a, b, and c into the formula. $$x = \dfrac{-(\square) \pm \sqrt{(\square)^2 - 4(\square)(\square)}}{2(\square)}$$ Solution:	
2. Solve $x^2 - 5x - 24 = 0$ using the quadratic formula. Identify a, b, and c. Solution:	

Example	Notes
4. Solve $3x^2 + 6 = 4x$ using the quadratic formula.	
Solution:	
5. Solve $8x^2 - 4x + 1 = 0$ using the quadratic formula.	
Solution:	

Helpful Hints

For all quadratic equations in the form $ax^2 + bx + c = 0$, $x = \dfrac{-b \pm \sqrt{b^2 - 4ac}}{2a}$.

If the discriminant is positive and a perfect square, then there are two real, rational solutions to the quadratic equation. If it is positive, but not a perfect square, then there are two real, irrational solutions. If it is zero, then there is one real, rational solution. If it is negative, then there are no real solutions and two non-real complex solutions.

Concept Check
1. Do quadratic equations always have real solutions? Complex solutions? Explain.

Practice
Solve by using the quadratic formula.

2. $-4x^2 = 12x + 9$

3. $x^2 + 6x + 13 = 0$

4. $x^2 + x - 42 = 0$

5. $2x^2 - x = -9$

Complex Numbers and More Quadratic Equations
Topic 33.4 Solving Equations Quadratic in Form

Vocabulary
quadratic formula   •   standard form   •   factored   •   quadratic in form

1. An equation is _____ if a substitution can be made to get an equation in the form $au^2 + bu + c = 0$.

Step-by-Step Video Notes
Watch the Step-by-Step Video lesson and complete the examples below.

Example	Notes
2 & 3. Find a substitution that will make the equation quadratic, and write the resulting quadratic equation. $2(x+4)^2 + 6(x+4) + 1 = 0$ $n^{-1/3} + 2n^{-1/6} - 3 = 0$	
4. Solve $x^4 - 10x^2 + 25 = 0$. Use the substitution $u = \boxed{}$. Rewrite the original equation. $x^4 - 10x^2 + 25 = 0$ $\boxed{} - \boxed{} + 25 = 0$ Solve by factoring. Solution:	

Example	Notes
5. Solve $6(x-1)^2 + (x-1) - 2 = 0$. Solution:	
6. Solve $x - 5x^{1/2} - 36 = 0$. Solution:	

Helpful Hints

If the quadratic expression in a quadratic equation will not factor, then the equation cannot be solved by factoring. Instead, another technique must be used, such as completing the square or the quadratic formula.

For all quadratic equations in the form $ax^2 + bx + c = 0$, $x = \dfrac{-b \pm \sqrt{b^2 - 4ac}}{2a}$.

Concept Check

1. Is the equation $x^9 + x^3 + 4 = 0$ quadratic in form? Is the equation $x^6 + x^3 + 4 = 0$ quadratic in form?

Practice

Find a substitution that will make the equation quadratic. Then, solve the equation.

2. $x^6 - 7x^3 - 8 = 0$

4. $15x - 7x^{1/2} - 2 = 0$

3. $(x-3)^2 - 6(x-3) + 9 = 0$

5. $x^4 - 3x^2 - 4 = 0$

Name: _____   Date: _____

Instructor: _____   Section: _____

Complex Numbers and More Quadratic Equations
Topic 33.5 Complex and Quadratic Applications

Vocabulary
Square Root Property • quadratic formula • surface area • area • resistance

1. To solve a quadratic equation in standard form $ax^2 + bx + c = 0$ using the _____,

   use $x = \dfrac{-b \pm \sqrt{b^2 - 4ac}}{2a}$.

Step-by-Step Video Notes
Watch the Step-by-Step Video lesson and complete the examples below.

Example	Notes
1. The surface area of a sphere is given by the formula $A = 4\pi r^2$. Solve for r. You do not need to rationalize the denominator. Answer:	
2. Solve $A = P(1+n)^2$ for n. Answer:	
3. A sail with the shape of a right triangle has an area of 39 square meters. If the boom, which forms the base of the sail, is 7 meters shorter than the mast, which forms the height, find the dimensions of the sail. Answer:	

Example	Notes
4. Find the resistance in a circuit where the voltage is $3 + 2i$ volts and the current is $3i$ amperes. Answer:	

Helpful Hints

The area of a triangle is $A = \dfrac{1}{2}bh$, where b is the base of the triangle, and h is the height.

The resistance R in an electrical circuit is given by the formula $R = \dfrac{V}{I}$, where V is the voltage, measured in volts, and I is the current in amperes.

Concept Check

1. The surface area of a cylinder is given by the formula $A = 2\pi r^2 + 2\pi rh$. Solve for r using the quadratic formula. You do not need to rationalize the denominator.

Practice

2. The area of a circle is given by the formula $A = \pi r^2$. Solve for r. You do not need to rationalize the denominator.

3. Solve $D = S(x-a)^2$ for x.

4. Heather has a sculpture on her shelf. The visible side is in the shape of a right triangle. The area of the sculpture is 30 square inches. If the height of the sculpture is 7 more inches than the base, what are the dimensions of the sculpture?

5. Find the resistance in a circuit where the voltage is $5 + i$ volts and the current is $4i$ amperes.

Name: _____   Date: _____

Instructor: _____   Section: _____

Graphing Quadratic Functions
Topic 34.1 Introduction to Graphing Quadratic Functions

Vocabulary
vertex   •   axis of symmetry   •   vertical shift of a parabola
horizontal shift of a parabola   •   vertex form of a quadratic equation

1. The _____ of a parabola is the vertical line through the vertex.

2. The _____ is $y = a(x - h)^2 + k$, where the vertex is (h, k).

3. The _____ of a parabola is the lowest or highest point on a parabola.

Step-by-Step Video Notes
Watch the Step-by-Step Video lesson and complete the examples below.

Example	Notes
1 & 2. Find the coordinates of the vertex of the parabola. $y = x^2 + 5$ $y = x^2 - 3$	
3 & 4. Find the coordinates of the vertex of the parabola. $y = (x + 6)^2$ $y = (x - 4.5)^2$	
5 & 6. Find the coordinates of the vertex of the parabola. $y = (x + 3)^2 + 2.5$ $y = (x - 8)^2 - 9$	

Example	Notes
7 & 8. Tell whether the parabola opens upward or downward, and whether it is wider or narrower than $y = x^2$. $y = \dfrac{2}{3}x^2$ $f(x) = -\dfrac{8}{5}(x-3)^2 + 7$	

Helpful Hints

The graph of $y = x^2 + k$ is a parabola shifted vertically k units from the origin. If $k > 0$, the parabola is shifted up. If $k < 0$, the parabola is shifted down. The vertex of the parabola is $(0, k)$.

The graph of $y = (x - h)^2$ is a parabola shifted horizontally h units from the origin. The vertex is $(h, 0)$. The horizontal shift of a parabola goes in the opposite direction of the sign of h. If the function is $(x - h)^2$, the parabola shifts right. If the function is $(x + h)^2$, the parabola shifts left.

Concept Check

1. How many units and in what direction does the graph of $y = 2(x+9)^2$ shift horizontally and vertically from $y = x^2$?

Practice

2. Find the coordinates of the vertex of the parabola.

    $f(x) = x^2 - 1$

3. Find the coordinates of the vertex of the parabola.

    $y = (x+15)^2$

4. Find the coordinates of the vertex of the parabola.

    $y = (x-2)^2 - 2.34$

5. Tell whether the parabola opens upward or downward, and whether it is wider or narrower than $y = x^2$.

    $g(x) = -\dfrac{1}{4}(x+5.2)^2 - 7$

418

Graphing Quadratic Functions
Topic 34.2 Finding the Vertex of a Quadratic Function

Vocabulary
vertex   •   vertex formula   •   quadratic function

1. The _____ of a parabola is the lowest or highest point on a parabola.

Step-by-Step Video Notes
Watch the Step-by-Step Video lesson and complete the examples below.

Example	Notes
1. Find the vertex of the quadratic function. Is the vertex a maximum or a minimum? $f(x) = x^2 - 8x + 15$ Identify a, b, and c.	
Find $x = \dfrac{-b}{2a}$.	
Substitute this x-value into the equation and find the value of the vertex.	
Determine if the vertex is a maximum or minimum.	
Answer:	

Example	Notes
2. Find the vertex of the quadratic function. Is the vertex a maximum or a minimum? $f(x) = 12x - 3x^2 - 6$ Answer:	
3. Find the vertex of the quadratic function. Is the vertex a maximum or a minimum? $f(x) = 5x^2 - 7$ Answer:	

Helpful Hints

The vertex of a parabola will occur at $x = \dfrac{-b}{2a}$.

If $a > 0$, the vertex is a minimum, and if $a < 0$, the vertex is a maximum.

Concept Check

1. Without finding the vertex of $f(x) = -(x-5)^2$, determine if it is a maximum or a minimum.

Practice

Find the vertex of the quadratic function. Is the vertex a maximum or a minimum?

2. $f(x) = x^2 + 4x - 10$

4. $f(x) = 5x + 3x^2 - 4$

3. $f(x) = -2x^2 - 16x - 12$

5. $f(x) = 1 - x^2$

Graphing Quadratic Functions
Topic 34.3 Finding the Intercepts of a Quadratic Function

Vocabulary
intercept   • x-intercept   • y-intercept   •   quadratic function

1. To find the _____ of a quadratic function, let $x = 0$ and evaluate $f(0)$.

2. To find the _____(s) of a quadratic function (if they exist), let $f(x) = 0$ and solve the equation for x.

Step-by-Step Video Notes
Watch the Step-by-Step Video lesson and complete the examples below.

Example	Notes
1 & 2. Find the y-intercept of each quadratic function. $f(x) = x^2 + 6x + 15$ $f(x) = 3x^2$	
3 & 4. Find the x-intercept(s) of each quadratic function, if any exist. $f(x) = x^2 + 2x - 24$ $f(x) = x^2 + 1$	

Example	Notes
5. Find the x- and y-intercepts of the quadratic function. $f(x) = x^2 + 5x - 14$ Answer:	
6. Find the x- and y-intercepts of the quadratic function. $f(x) = x^2 - 6x + 5$ Answer:	

Helpful Hints
The graph of a quadratic equation or function may have zero, one, or two x-intercepts. However, the graph will always have exactly one y-intercept.

For a quadratic function, $f(x) = ax^2 + bx + c$, the y-intercept is always the point $(0, c)$.

Concept Check
1. Why must a quadratic function have no more than one y-intercept?

Practice
Find the x- and y-intercepts of each quadratic function.

2. $f(x) = x^2 - 6x - 16$ 4. $f(x) = x^2 + 11$

3. $f(x) = 3x^2 + 18x + 27$ 5. $f(x) = x^2 - 9x + 20$

Name: _____   Date: _____

Instructor: _____   Section: _____

Graphing Quadratic Functions
Topic 34.4 Graphing Quadratic Functions Summary

Vocabulary
x-intercept • *y*-intercept • axis of symmetry • parabola • vertex

1. A parabola is symmetric over the _____, which means that the graph looks the same on both sides of this line.

Step-by-Step Video Notes
Watch the Step-by-Step Video lesson and complete the examples below.

Example	Notes
1. Graph the function $f(x) = x^2$. Evaluate the function for different values. Then plot the points and draw a smooth curve through them. 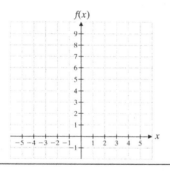	
2. Graph the function $f(x) = x^2 - 6x + 8$. 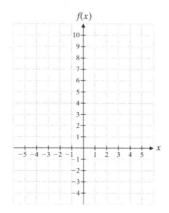	

Example	**Notes**

3. Graph the function $f(x) = -2x^2 + 4x - 3$.

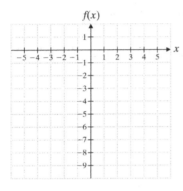

Helpful Hints
When graphing a quadratic equation, determine if the parabola opens upward or downward, making the vertex either a minimum or a maximum, respectively. Identify and plot the vertex, the x-intercept(s), the y-intercept, and other ordered pairs that satisfy the function, if needed. Draw a smooth curve through the points to form the parabola.

Concept Check
1. For examples 1 through 3, find and graph the axis of symmetry.

Practice
Graph each function.

2. $f(x) = x^2 - 2x - 3$

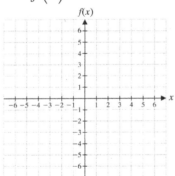

3. $f(x) = -2x^2 + 2$

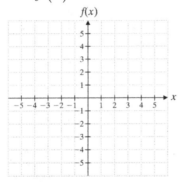

4. $f(x) = -x^2 + 4x - 5$

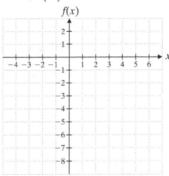

Graphing Quadratic Functions
Topic 34.5 Applications with Quadratic Functions

Vocabulary

optimization   •   vertex   •   maximum   •   minimum

1. _____ is the process of finding the maximum or minimum value of a function.

Step-by-Step Video Notes
Watch the Step-by-Step Video lesson and complete the examples below.

Example	Notes
1. The cost C in dollars of producing x bicycles at a factory can be modeled by the function $C(x) = 2x^2 - 760x + 85,000$. Find the number of bicycles that must be manufactured to minimize the cost. Then find the minimum cost. Answer:	
2. A roll of chain-link fence is 500 feet long. What are the maximum dimensions of a rectangular field that can be enclosed with this fence? What is the maximum area of this field? Answer:	

Example	Notes
3. A rocket is fired upward with an initial velocity v of 110 meters per second. The quadratic function $h(t) = -5t^2 + 110t$ can be used to find the height $h(t)$ of the rocket, in meters, at any time t in seconds. Find the maximum height of the rocket and when it occurs.	

Answer:

Helpful Hints

If $a > 0$, the parabola opens upward and the vertex is the lowest point (or minimum). If $a < 0$, the parabola opens downward and the vertex is the highest point (or maximum).

To find the maximum or minimum value of a parabola, find the y-coordinate of the vertex.

Concept Check

1. You are asked to minimize the cost function $C(x) = -x^2 + 500x + 75,000$. Explain why this is not possible.

Practice

The following cost functions model the cost C, in dollars, of producing x items. For each function, find the number of items that must be manufactured to minimize the cost. Then find the minimum cost.

The following functions give the height $h(t)$ of an object that is projected upward, in meters, at any time t in seconds. Find the maximum height of the object and when it occurs.

2. $C(x) = 5x^2 - 980x + 63,000$

4. $h(t) = -5t^2 + 300t$

3. $C(x) = 3x^2 - 804x + 94,000$

5. $h(t) = -5t^2 + 8t + 100$

Compound and Nonlinear Inequalities
Topic 35.1 Interval Notation

Vocabulary

graph of an inequality • infinity • negative infinity • interval notation • union

1. A(n) _____ is a picture that represents all of the solutions of the inequality.

2. The _____ of two sets is the set of every element that is in either or both sets.

Step-by-Step Video Notes

Watch the Step-by-Step Video lesson and complete the examples below.

Example	Notes
1 & 2. Graph each inequality and write in interval notation. $x < -2$ $0 \le x$	
3. Graph each inequality and write in interval notation. $-1 \le x < 4$	
4. Graph each inequality and write in interval notation. $-3 \le x \le 0$	

Example	Notes

7 & 8. Graph each inequality and write in interval notation.

$x < -3$ or $x > 2$

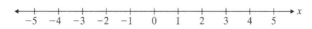

$x \leq -2$ or $x \geq 0$

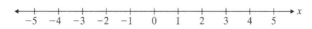

Helpful Hints

Interval notation is another way to represent the solution to an inequality. It uses two values, the starting point and the ending point of the solution.

In interval notation, a parenthesis is used for endpoints that are not included in the interval and a bracket is used for endpoints that are included in the interval.

Do not confuse interval notation with an ordered pair.

When the variable is written first, or on the left in an inequality, the arrow on the graph points in the same direction as the inequality sign.

Concept Check

1. Describe in words what numbers are included in the interval $[-4.5, \infty)$.

Practice

Graph each inequality and write in interval notation.

2. $x \geq 3$

Write the interval notation for the given graphs of inequalities.

4.

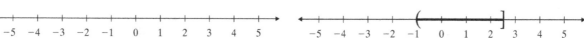

3. $0 < x \leq 3$

5.

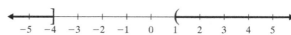

Compound and Nonlinear Inequalities
Topic 35.2 Graphing Compound Inequalities

Vocabulary
compound inequality   •   graph of a compound inequality

1. A _____ is two or more inequalities separated by "and" or "or."

Step-by-Step Video Notes
Watch the Step-by-Step Video lesson and complete the examples below.

Example	Notes
1. Graph. Write your answer in interval notation. $x > -2$ and $x \leq 4$ $\begin{array}{c} \text{—} + + + + + + + + + + \text{→} \\ -5\ -4\ -3\ -2\ -1\ \ 0\ \ 1\ \ 2\ \ 3\ \ 4\ \ 5 \end{array}$	
2–4. Graph. $-3 \leq x < 3$ $\begin{array}{c} \text{—} + + + + + + + + + + \text{→} \\ -5\ -4\ -3\ -2\ -1\ \ 0\ \ 1\ \ 2\ \ 3\ \ 4\ \ 5 \end{array}$ $(-2, 5]$ $\begin{array}{c} \text{—} + + + + + + + + + + \text{→} \\ -5\ -4\ -3\ -2\ -1\ \ 0\ \ 1\ \ 2\ \ 3\ \ 4\ \ 5 \end{array}$ $x > 4$ and $x < 1$ $\begin{array}{c} \text{—} + + + + + + + + + + \text{→} \\ -5\ -4\ -3\ -2\ -1\ \ 0\ \ 1\ \ 2\ \ 3\ \ 4\ \ 5 \end{array}$	

Example	Notes
6 & 7. Graph. Write your answer in interval notation. $x > 3$ or $x \leq -2$  $x \geq -1$ or $x \leq 5$ 	

9. Graph. Write your answer in interval notation.

$x < 5$ or $x < -3$

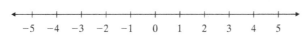

Helpful Hints

The solution to a compound inequality using the word "and" is the set of all the points on a number line that satisfy both the inequalities. The interval of the graph is the intersection of the intervals of each part of the inequality.

The solution to a compound inequality using the word "or" is the set of all the numbers on a number line that satisfy either one of the inequalities. The interval of the graph is the union of the intervals of each part of the inequality.

Concept Check
1. Can you give examples of any numbers not in the solution set of $x \leq 0$ or $x > -5$?

Practice
Graph. Write your answer in interval notation.

2. $y \geq -3$ and $y < 5$

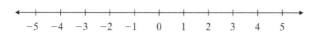

3. $-5 < x < 1$

4. $x < 2$ or $x > 2$

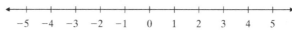

5. $x > -1.5$ or $x \leq 2$

Compound and Nonlinear Inequalities
Topic 35.3 Solving Compound Inequalities

Vocabulary
compound inequality   •   interval notation   •   inequality

1. To solve a(n) _____, solve each of the inequalities for the variable and express the solution in _____.

Step-by-Step Video Notes
Watch the Step-by-Step Video lesson and complete the examples below.

Example	Notes
1. Solve and graph $x + 1 > -2$ and $x - 2 \leq 2$. Then write the solution in interval notation. 	
2. Solve $3x + 2 \geq 14$ or $2x - 1 \leq -7$ for x and graph the solution. Then write the solution using interval notation. 	

431

Example	Notes
4. Solve $6+x<9$ or $5x+2>-18$ for x and graph the solution. Then write the solution using interval notation. 	
5. Solve $-9 \leq 3a-3 < 9$ for a and graph the solution. Then write the solution using interval notation. 	

Helpful Hints

To solve an inequality, use the same procedure used to solve equations, except reverse the direction of the inequality if you multiply or divide by a negative number.

Concept Check

1. Write the two separate inequalities you must solve to find the solution to the compound inequality $-2 \leq -4x+6 < 14$.

Practice

Solve for x and graph the solution. Then write the solution using interval notation.

2. $x+2 \geq -1$ and $x+7 \leq 9$ 4. $-7x+8 < -27$ and $3x+10 \leq 4$

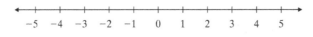 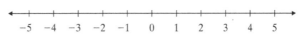

3. $2x-7 < 3$ and $6x+4 \leq -2$ 5. $-8 \leq 3m-14 \leq 1$

Compound and Nonlinear Inequalities
Topic 35.4 Solving Quadratic Inequalities

Vocabulary

quadratic inequality   •   compound inequality   •   quadratic formula

1. To solve a _____, replace the inequality symbol with an equal sign. Solve the resulting equation to find the solutions, or boundary numbers. Use these numbers to separate the number line into regions to test the original for solutions.

Step-by-Step Video Notes
Watch the Step-by-Step Video lesson and complete the examples below.

Example	Notes
1. Solve $(x-3)(x+1)>0$ and graph the solution. Write the solution in interval notation.	

| 2. Solve $2x^2 + x - 6 \leq 0$ and graph the solution. Write the solution in interval notation. | |

Example	Notes

3. Solve $x^2 - 4x + 4 > 0$ and graph the solution. Write the solution in interval notation.

4. Solve $x^2 + 4x - 6 \geq 0$ and graph the solution. Write the solution in interval notation.

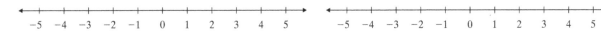

Helpful Hints
When testing to see which regions of the number line satisfy the original inequality, choose numbers that make the calculations easier. Integers are generally easier to use in calculations than decimals or fractions.

Concept Check
1. Is there a quadratic inequality for which the solution would be all real numbers? No solution?

Practice
Solve and graph. Then write the solution in interval notation.

2. $x^2 - 2x - 15 > 0$ 4. $x^2 + 3x - 9 \leq 0$

3. $5x^2 - 14x - 3 < 0$ 5. $x^2 - 10x + 25 \geq 0$

Name: _____   Date: _____

Instructor: _____   Section: _____

Compound and Nonlinear Inequalities
Topic 35.5 Solving Rational Inequalities

Vocabulary
rational inequality   •   quadratic inequality   •   boundary number

1. A _____ is an inequality that contains one or more rational expressions.

Step-by-Step Video Notes
Watch the Step-by-Step Video lesson and complete the examples below.

Example	Notes
1. Solve and graph the solution. Write the solution in interval notation. $$\frac{(x-3)}{(x+1)} < 0$$ 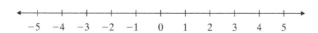	
2. Solve and graph the solution. Write the solution in interval notation. $$\frac{x^2 + 8x + 15}{x - 1} \leq 0$$ 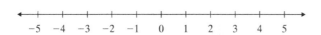	

Example	Notes
3. Solve and graph the solution. Write the solution in interval notation. $$\frac{x+9}{x}+2\le 0$$ 	

Helpful Hints

The first step in solving a rational inequality is to solve the inequality to have 0 on one side, and a single expression on the other side.

If a rational inequality uses < or >, then the boundary points are not included in the solution interval. If the rational inequality uses ≤ or ≥, then the boundary points are included in the solution interval, except for any boundary points that would make the denominator equal to zero.

Concept Check

1. Without solving $\dfrac{x^2+4x-12}{x^2-1}\ge 0$, determine the boundary points and state whether each boundary point is included or not included in the solution interval.

Practice

Solve and graph the solution. Write the solution in interval notation.

2. $\dfrac{x-4}{x+2}>0$

4. $\dfrac{5x-12}{x}\ge -1$

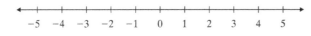

3. $\dfrac{x^2-x-12}{x+5}<0$

5. $\dfrac{x^2+2x-3}{x^2-2x-8}\le 0$

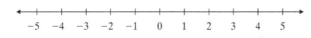

Absolute Value Equations and Inequalities
Topic 36.1 Introduction to Absolute Value Equations

Vocabulary
absolute value   •   absolute value equation rule

1. The _____ states that if $|x|=a$, and a is a positive real number,
   then $x=a$ or $x=-a$.

Step-by-Step Video Notes
Watch the Step-by-Step Video lesson and complete the examples below.

Example	Notes
2–4. Solve. $|x|=8.35$ $|y|=0$ $|x|=-11$	
5. Solve $|x|+1=10$. Isolate the absolute value expression. Write as two equations according to the absolute value equation rule. Solve by applying the rule for absolute value equations. Answer:	

Example	Notes
6. Solve $-6|x| = -72$.	
Answer:	

7 & 8. Write an equation using absolute value to represent the given graph.

Helpful Hints

The absolute value of a number is the distance between that number and zero on a number line.

Concept Check

1. Why does the equation $|x| = -6$ have no solution?

Practice

Solve.

2. $|n| = 6\frac{7}{8}$

4. $\frac{1}{3}|a| = 6$

3. $|x| + 5.8 = 3.2$

5. $-4|x| = -5.2$

Absolute Value Equations and Inequalities
Topic 36.2 Solving Basic Absolute Value Equations

Vocabulary
absolute values   •   absolute value equations rule

1. The first step to solving an absolute value equation is to isolate the _____.

Step-by-Step Video Notes
Watch the Step-by-Step Video lesson and complete the examples below.

Example	Notes
1. Solve $\lvert x+1 \rvert = 6$. Answer:	
2. Solve $\lvert 5y \rvert = 35$. Answer:	
3. Solve $\left\lvert \dfrac{1}{2}x - 1 \right\rvert = 5$. Write the absolute value equation as two separate equations. Answer:	

Example	Notes
4. Solve $\lvert 3x-1 \rvert + 2 = 5$. Answer:	
5. Solve $\lvert 8x-3 \rvert + 12 = 7$. Answer:	

Helpful Hints

Remember, the absolute value of a number is the distance between that number and zero on a number line.

If $\lvert x \rvert = a$, and a is a positive real number, then $x = a$ or $x = -a$.

Concept Check

1. If $\lvert 7x - 3y \rvert = 4z$, then $7x - 3y$ is equal to what two numbers?

Practice

Solve.

2. $\lvert x + 7 \rvert = 13$

3. $\lvert 9x - 6 \rvert = 12$

4. $\left\lvert \dfrac{2}{3} - \dfrac{1}{6}x \right\rvert = 3$

5. $\lvert 2x - 1 \rvert - 5 = 4$

Absolute Value Equations and Inequalities
Topic 36.3 Solving Multiple Absolute Value Equations

Vocabulary
absolute value expressions   •   multiple absolute value equations

1. In order for two _____ to be equal, the expressions inside the
    absolute value bars must be either equal to each other or opposites of each other.

Step-by-Step Video Notes
Watch the Step-by-Step Video lesson and complete the examples below.

Example	Notes
1. Solve $\lvert 3x + 4 \rvert = \lvert x \rvert$. Write the absolute value equation as two separate equations. Solve the resulting equations. Answer:	
2. Solve $\lvert 3x - 4 \rvert = \lvert x + 6 \rvert$. Write the absolute value equation as two separate equations. Answer:	

Example	Notes
3. Solve $\lvert x+3 \rvert = \lvert x-5 \rvert$.	
Answer:	
4. Solve $\lvert 2x-4 \rvert = \lvert 4-2x \rvert$.	
Answer:	

Helpful Hints

When solving equations with an absolute value expression alone on both sides of the equation, if the expressions inside the absolute value bars are equivalent or opposites, the equation is an identity, and all real numbers are solutions of the equation.

Concept Check

1. Marissa reasons that since $\lvert x \rvert = \lvert y \rvert$, then $x = y$. Is she correct? Explain.

Practice

Solve.

2. $\lvert 4x+12 \rvert = \lvert 2x \rvert$

4. $\lvert 4-x \rvert = \left\lvert \dfrac{x}{3}+2 \right\rvert$

3. $\lvert x-4 \rvert = \lvert 4x+11 \rvert$

5. $\lvert 3x+2 \rvert = \lvert 3x-8 \rvert$

Absolute Value Equations and Inequalities
Topic 36.4 Solving Absolute Value Inequalities

Vocabulary

$|x| < a$ • $|x| > a$ •   absolute value equation rule

1. If a is a positive real number, and _____, then $-a < x < a$ or $x > -a$ and $x < a$.

Step-by-Step Video Notes
Watch the Step-by-Step Video lesson and complete the examples below.

Example	Notes
1 & 2. Graph. Write the answer in interval notation.	

$|x| \geq 2$

$|x| < 3$

3. Solve and graph $|x + 5| \leq 10$. Write the answer in interval notation.	

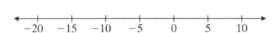

Answer:

Example	Notes
4. Solve and graph $\|-2x-1\| \geq 7$. Write the answer in interval notation. 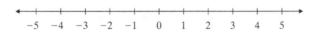 Answer:	
5. Solve and graph $\|4x+2\|+5 > 9$. Write the answer in interval notation. 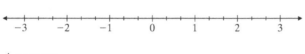 Answer:	

Helpful Hints

If $\|x\| > a$ and a is a positive real number, then $x < -a$ or $x > a$. The interval notation for the solution is $(-\infty, a) \cup (a, \infty)$.

When graphing inequalities, use an open circle (or parenthesis) for $<$ or $>$ and a closed circle (or bracket) for $\leq$ or $\geq$.

Concept Check

1. Sketch the number line graph that displays the interval $(-\infty, -3] \cup [6, \infty)$.

Practice

Solve and graph. Write the answer in interval notation.

2. $\|x+5\| \leq 4$ 　　　　　　　　　　　　4. $\|-4x+3\| \geq 11$

3. $\|3x-2\| < 7$ 　　　　　　　　　　　　5. $\|4x-7\|+8 > 17$

444

Conic Sections
Topic 37.1 Introduction to Conic Sections

Vocabulary
conic section • circle • parabola • ellipse • hyperbola

1. In general, a(n) _____ is a shape that can be formed by slicing a cone with a plane.

Step-by-Step Video Notes
Watch the Step-by-Step Video lesson and complete the examples below.

Example	**Notes**
1 & 2. Identify the conic sections shown.	

1 & 2. Identify the conic sections shown.

5–8. Identify the conic sections described as a circle, a parabola, an ellipse, or a hyperbola.

For each point in the set, the absolute value of the difference of its distances to two fixed points is constant.

For each point in the set, the sum of its distances to two fixed points is constant.

Each point in the set is a fixed distance from a center point.

This is the graph of a quadratic equation.

Example	Notes
9–12. Identify each conic section from its equation as a circle, a parabola, an ellipse, or a hyperbola. $x = -3(y-5)^2 + 7$ $$\dfrac{(y-2)^2}{3^2} - \dfrac{(x+6)^2}{11^2} = 1$$ $$\dfrac{(y+3)^2}{1^2} + \dfrac{(x+8)^2}{9^2} = 1$$ $$(x+2)^2 + (y-6)^2 = 5^2$$	

Helpful Hints

In a given plane, a circle is the set of all points that are a fixed distance from a center point. A parabola is the graph of a quadratic equation.

An ellipse is the set of points in a plane such that for each point in the set, the sum of its distances to two fixed points is constant. A hyperbola is the set of points in a plane such that for each point in the set, the absolute value of the difference of its distances to two fixed points is a constant.

Concept Check

1. Identify the difference between the equation of an ellipse centered at the origin and the equation of a hyperbola centered at the origin.

Practice

Identify each conic section from its equation as a circle, a parabola, an ellipse, or a hyperbola.

2. $y = 3(x+1)^2 - 2$

3. $x^2 + (y+3)^2 = 14^2$

4. $\dfrac{y^2}{4.8^2} - \dfrac{x^2}{1.1^2} = 1$

5. Identify the conic section shown.

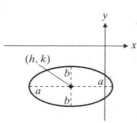

Conic Sections
Topic 37.2 The Circle

Vocabulary
conic section   •   circle   •   standard form

1. A _____ is a shape that can be formed by slicing a cone with a plane.

2. In a given plane, a _____ is a set of all points that are a fixed distance from a center point.

Step-by-Step Video Notes
Watch the Step-by-Step Video lesson and complete the examples below.

Example	Notes
2. Find the center and radius of the following circle. Then sketch its graph. $$(x+2)^2+(y-3)^2=25$$ Recall the standard form of the equation of a circle, $(x-h)^2+(y-k)^2=r^2$. Find the center. Find the radius. Sketch the graph.	

4. Write the equation of the circle in standard form with the given center and radius. Center $(8,-2)$; $r=7$ Answer:	

Example	Notes
6. Write the equation of the circle in standard form. Find the center and the radius of the circle.	

$$x^2 + 6x + y^2 - 8y + 7 = 0$$

$$x^2 + 6x + \boxed{} + y^2 - 8y + \boxed{} = -\boxed{} + \boxed{} + \boxed{}$$

$$\left(x + \boxed{}\right)^2 + \left(y - \boxed{}\right)^2 = \boxed{}$$

Answer:

7. Write the equation of the circle in standard form. Find the radius and center of the circle and sketch its graph.

$$x^2 + 2x + y^2 + 6y + 6 = 0$$

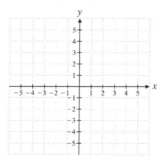

Example	Notes
8. A Ferris wheel has a radius, r, of 25.3 feet. The height of the tower, t, is 31.8 feet. The distance, d, from the origin to the base of the tower is 44.8 feet. Find the standard form of the equation of the circle represented by the Ferris wheel.	

Answer:

Helpful Hints

The standard form of the equation of a circle with center at (h,k) and radius r is

$$(x-h)^2 + (y-k)^2 = r^2.$$

Be careful of the signs when making calculations with h and k. For example, the equation $(x+2)^2 + (y-3)^2 = 25$ has a center of $(-2,3)$, not $(2,3)$.

Concept Check
1. Explain why a circle can have negative coordinates for its center but cannot have a negative radius.

Practice
For the circle $x^2 - 2x + y^2 + 2y = 2$, determine the following information.

2. Write the equation of the circle in standard form.

4. Graph the circle and label the center and radius.

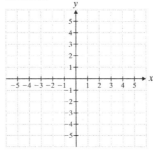

3. Identify the center and radius of the circle.

Additional Notes

Conic Sections
Topic 37.3 The Parabola

Vocabulary

conic section   •   parabola   •   vertical parabola

horizontal parabola   •   vertex   •   axis of symmetry

1. A(n) _____ is the graph of a quadratic equation.

2. The _____ is either the highest or lowest point on the parabola.

Step-by-Step Video Notes
Watch the Step-by-Step Video lesson and complete the examples below.

Example	Notes
1. Graph the parabola $y = 2(x-1)^2 - 3$.	

Determine if the parabola is vertical or horizontal.

Identify a, h, and k.

Determine the direction the parabola opens.

Find the vertex.

Find the axis of symmetry.

Find the y-intercept.

Graph the parabola.

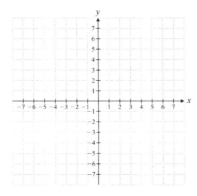

Example	Notes
2. Graph the parabola $x = (y+2)^2 - 3$. 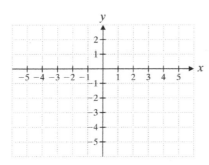	

Helpful Hints

The standard form of the equation of a vertical parabola is $y = a(x-h)^2 + k$, where $a \neq 0$. The parabola opens upward if $a > 0$ and downward if $a < 0$. The vertex of the parabola is (h, k) and the axis of symmetry is the line $x = h$.

The standard form of the equation of a horizontal parabola is $x = a(y-k)^2 + h$, where $a \neq 0$. The parabola opens to the right if $a > 0$ and to the left if $a < 0$. The vertex of the parabola is (h, k) and the axis of symmetry is the line $y = k$.

The symmetry of a parabola can be used to find points on the parabola. If we know a point on the parabola (aside from the vertex), then we can find another by reflecting (or folding) it over the axis of symmetry.

Concept Check

1. Do parabolas always have both x- and y-intercepts? Explain.

Practice

Graph the parabolas. Label the vertices and axes of symmetry.

2. $y = (x-2)^2 + 3$

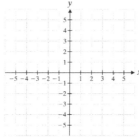

3. $y = -(x+3)^2 - 1$

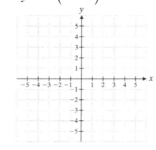

4. $x = -3(y+3)^2 - 2$

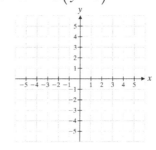

Name: _____   Date: _____

Instructor: _____   Section: _____

Conic Sections
Topic 37.4 The Ellipse

Vocabulary

conic section   •   ellipse   •   focus

1. A(n) _____ is the set of points in a plane such that for each point in the set, the sum of its distances to two fixed points is constant.

Step-by-Step Video Notes
Watch the Step-by-Step Video lesson and complete the examples below.

Example	Notes
2. Graph the ellipse. Label the intercepts. $$x^2 + 3y^2 = 12$$ 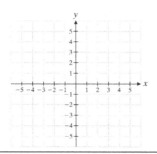	
3. Graph the ellipse $\dfrac{(x-5)^2}{4} + \dfrac{(y-6)^2}{9} = 1$. 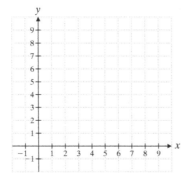	

Example	Notes
4. The orbit of Venus is an ellipse with the Sun as a focus. If we say that the center of the ellipse is at the origin, an approximate equation for the orbit is $\dfrac{x^2}{5013} + \dfrac{y^2}{4970} = 1,$ where x and y are measured in millions of miles. Find the largest possible distance across the ellipse. Round your answer to the nearest million miles.	

Answer:

Helpful Hints

An ellipse centered at the origin has the equation $\dfrac{x^2}{a^2} + \dfrac{y^2}{b^2} = 1,$ where $a > 0$ and $b > 0$. The intercepts of this ellipse are at $(a,0)$, $(-a,0)$, $(b,0)$, and $(-b,0)$.

An ellipse with center at (h,k) has the equation $\dfrac{(x-h)^2}{a^2} + \dfrac{(y-k)^2}{b^2} = 1,$ where $a > 0$ and $b > 0$.

Concept Check
1. Explain why every circle is an ellipse but not every ellipse is a circle.

Practice
Graph the ellipses. Label the vertices.

2. $\dfrac{x^2}{16} + \dfrac{y^2}{25} = 1$

3. $\dfrac{(x+1)^2}{9} + \dfrac{(y-2)^2}{4} = 1$

4. $x^2 + \dfrac{(y+3)^2}{4} = 1$

Conic Sections
Topic 37.5 The Hyperbola

Vocabulary
conic section   •   hyperbola   •   asymptotes of hyperbolas
fundamental rectangle of a hyperbola

1. A(n) _____ is the set of points in a plane such that for each point in the set, the absolute value of the difference of its distances to two fixed points (called foci) is a constant.

Step-by-Step Video Notes
Watch the Step-by-Step Video lesson and complete the examples below.

Example	Notes
1. Graph the hyperbola $\dfrac{x^2}{25} - \dfrac{y^2}{16} = 1$. 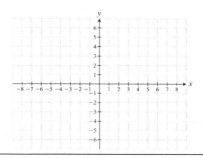	
2. Graph the hyperbola $4y^2 - 7x^2 = 28$. 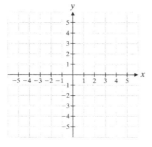	

Example	Notes
3. Graph the hyperbola $\dfrac{(x-4)^2}{9} - \dfrac{(y-5)^2}{4} = 1.$ 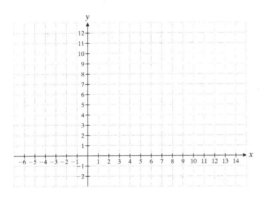	

Helpful Hints

Let a and b be any positive real numbers. A horizontal hyperbola with center at (h,k) and vertices $(h-a,k)$ and $(h+a,k)$ has the equation $\dfrac{(x-h)^2}{a^2} - \dfrac{(y-k)^2}{b^2} = 1.$

Let a and b be any positive real numbers. A vertical hyperbola with center at (h,k) and vertices $(h,k-b)$ and $(h,k+b)$ has the equation $\dfrac{(y-k)^2}{b^2} - \dfrac{(x-h)^2}{a^2} = 1.$

Concept Check
1. Compare and contrast the four types of conic sections (circles, parabolas, ellipses, and hyperbolas).

Practice
Graph the hyperbolas.

2. $9x^2 - 5y^2 = 45$

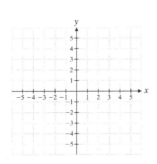

3. $-\dfrac{(x+1)^2}{9} + \dfrac{(y-2)^2}{4} = 1$

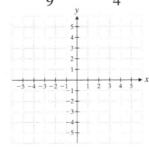

4. $\dfrac{(x-1)^2}{4} - y^2 = 1$

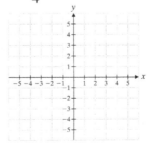

Logarithmic and Exponential Functions
Topic 38.1 Composite Functions

Vocabulary
function composition   •   composite functions   •   volume

1. _____ is a way of using the output of one function as the input of another function.

Step-by-Step Video Notes
Watch the Step-by-Step Video lesson and complete the examples below.

Example	Notes
1 & 2. Given $f(x) = 2x$ and $g(x) = 4x + 7$, find the following. $$f[g(x)] = f\left[\boxed{}\right]$$ $$= \boxed{}\left(\boxed{}\right)$$ $$g[f(x)] = g\left[\boxed{}\right]$$	
3 & 4. Given $f(x) = 4x^2$ and $g(x) = 3x - 5$, find the following. $(f \circ g)(x)$ $(g \circ f)(4)$	

Example	Notes
5. Given $f(x) = \dfrac{7}{2x-3}$ and $g(x) = x+2$, find $f[g(x)]$.	

Answer:

Helpful Hints

The composition of functions is not commutative. That is, for two functions $f(x)$ and $g(x)$, $f[g(x)]$ is not the same as $g[f(x)]$.

Be sure to state any excluded values when finding the composition of two functions. Values that make the denominator zero must be excluded from the domain because division by zero is undefined.

Concept Check

1. Given $f(x) = \dfrac{x}{2}$, find $g(x)$ such that $f[g(x)] = g[f(x)]$.

Practice

Given $f(x) = 2x-1$ and $g(x) = x^3 + 8$, find the following.

2. $f[g(x)]$

3. $g[f(2)]$

Given $f(x) = x^2$ and $g(x) = \dfrac{3x}{x+10}$, find the following.

4. $(f \circ g)(x)$

5. $(g \circ f)(0)$

Logarithmic and Exponential Functions
Topic 38.2 Inverse Functions

Vocabulary
function   •   one-to-one function   •   horizontal line test   •   inverse function

1. A(n) _____ is a function that reverses the domain and range of a one-to-one function.

2. A(n) _____ is a function for which every y value in the range has one and only one x value in the domain.

Step-by-Step Video Notes
Watch the Step-by-Step Video lesson and complete the examples below.

Example	Notes
1 & 2. Determine whether the relations are one-to-one functions. $$F = \{(1,-3),(2,0),(3,3),(4,6),(5,9)\}$$ $$G = \{(-2,5),(-1,2),(0,1),(1,2),(2,5)\}$$	
4. Determine whether the relation is a one-to-one function. 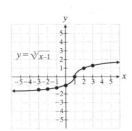	
5. Find the inverse of the one-to-one function. $$A = \{(1,-3),(2,0),(3,3),(4,6),(5,9)\}$$ Answer:	

Example	Notes
7. Find the inverse of the one-to-one function. $f(x) = x^3 + 7$ Answer:	
8. Graph $f(x) = 2x + 2,$ then find and graph its inverse. 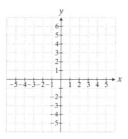	

Helpful Hints

The horizontal line test states that if a horizontal line can pass along the y-axis and cross the graph of a function in at most one place, then the graph represents a one-to-one function.

The graphs of a function and its inverse are symmetric about the line $y = x$.

Concept Check

1. Does $f(x) = x^2$ have an inverse function? Why?

Practice

2. Determine whether the relations are one-to-one functions.

$B = \{(-1,7),(0,9),(1,11),(2,13)\}$

$C = \{(-2,6),(0,0),(2,6),(3,12)\}$

3. Find the inverse of the one-to-one function.

$f(x) = 3x + 1$

4. Graph $f(x) = 2x + 4,$ then find and graph its inverse.

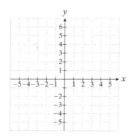

Logarithmic and Exponential Functions
Topic 38.3 Evaluating Exponential and Logarithmic Expressions

Vocabulary
logarithm   •   exponential expression

1. The _____, base b, of a positive number, x, is the power (exponent) to which the base b must be raised to produce x and is represented by the expression $\log_b x$.

Step-by-Step Video Notes
Watch the Step-by-Step Video lesson and complete the examples below.

Example	Notes
1–3. Evaluate each exponential expression for the given value of x. 5^x; $x = 2$ 4^x; $x = 3$ 2^x; $x = 5$	
5 & 6. Evaluate the exponential expression for the given value of x. 2^x; $x = -1$ 6^x; $x = -2$	

Example	Notes
7–10. Evaluate each logarithmic expression. $\log_2 4$ $\log_5 5$ $\log_3 \dfrac{1}{27}$ $\log_4 1$	

Helpful Hints

No matter the base b, $\log_b x$ for $0 < x < 1$ is always negative, whereas $\log_b x$ is always positive for $1 < x$.

No matter the base b, $\log_b 1 = 0$.

No matter the base b, $\log_b b = 1$.

Concept Check

1. If $4^{10} = 1{,}048{,}576$, what is $\log_4 1{,}048{,}576$?

Practice

Evaluate the exponential expression for the given values of x.

2. 9^x; $x = 3$, $x = 0$

3. 3^x; $x = -2$, $x = -4$

Evaluate each logarithmic expression.

4. $\log_4 64$

5. $\log_5 \dfrac{1}{25}$

6. $\log_{12{,}589.654} 1$

Logarithmic and Exponential Functions
Topic 38.4 Graphing Exponential Functions

Vocabulary
exponential function   • e •   base

1. A(n) _____ is a function of the form $f(x) = b^x$, where $b > 0$, $b \neq 1$, and x is a real number.

2. The number _____ is an irrational number that occurs in many formulas that describe real-world phenomena, such as the growth of cells and radioactive decay.

Step-by-Step Video Notes
Watch the Step-by-Step Video lesson and complete the examples below.

Example	**Notes**
1. Graph the exponential function $f(x) = 2^x$. 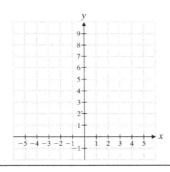	
2. Graph the exponential function $f(x) = \left(\dfrac{1}{2}\right)^x$. 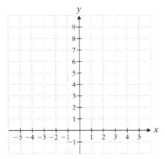	

Example	Notes
3. Graph the exponential function $f(x) = e^x$. 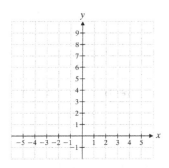	

Helpful Hints

An asymptote is a line that the graph of a function gets closer to as the value of x gets larger or smaller. The x-axis is an asymptote for every exponential function of the form $f(x) = b^x$.

The domain of an exponential function $f(x) = b^x$ is the set of all real numbers, whereas the range of an exponential function $f(x) = b^x$ is the set of all positive real numbers.

Concept Check

1. Explain why the graph of $f(x) = b^x$ always passes through $(0,1)$ and $(1,b)$.

Practice

Graph the exponential functions.

2. $f(x) = 3^x$

3. $f(x) = \left(\dfrac{1}{4}\right)^x$

4. $f(x) = \left(\dfrac{1}{e}\right)^x$

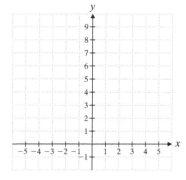
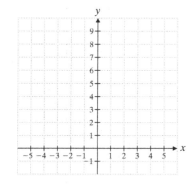
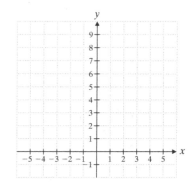

Logarithmic and Exponential Functions
Topic 38.5 Converting Between Exponential and Logarithmic Forms

Vocabulary
logarithmic equation   •   exponential equation   •   logarithm

1. The _____ $y = \log_b x$ is the same as the _____ $x = b^y$, where $b > 0$ and $b \neq 1$.

Step-by-Step Video Notes
Watch the Step-by-Step Video lesson and complete the examples below.

Example	Notes
1. Write $81 = 3^4$ in logarithmic form.	

2–4. Write in logarithmic form.

$49 = 7^2$

$512 = 8^3$

$\dfrac{1}{100} = 10^{-2}$

Example	Notes
5. Write $2 = \log_5 25$ in exponential form.	

6–8. Write in exponential form.

$$\frac{1}{2} = \log_{16} 4$$

$$0 = \log_{17} 1$$

$$-4 = \log_{10}\left(\frac{1}{10,000}\right)$$

Helpful Hints

The logarithm, base b, of a positive number, x, is the power (exponent) to which the base b must be raised to produce x and is represented by the expression $\log_b x$.

Be careful with signs when changing forms.

Concept Check

1. When converting $14^2 = 196$ to logarithmic form, Melissa wrote $196 = \log_2 14$. Explain her error.

Practice

Write in logarithmic form.

2. $13^2 = 169$

3. $\left(\frac{1}{7}\right)^4 = \frac{1}{2401}$

Write in exponential form.

4. $1 = \log_{4289} 4289$

5. $-3 = \log_6 \frac{1}{216}$

Logarithmic and Exponential Functions
Topic 38.6 Graphing Logarithmic Functions

Vocabulary
logarithmic function   •   exponential function   •   asymptote

1. A(n) _____ is a function of the form $f(x) = \log_b x$, where $b > 0$, $b \neq 1$, and
    x is a positive real number.

Step-by-Step Video Notes
Watch the Step-by-Step Video lesson and complete the examples below.

Example	Notes
1. Graph the logarithmic function $f(x) = \log_2 x$. 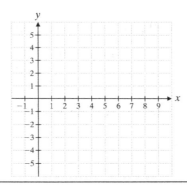	
2. Graph the logarithmic function $f(x) = \log_{1/3} x$. 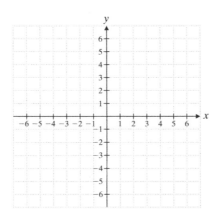	

Example	Notes
3. Graph $f(x) = \log_2 x$ and $f(x) = 2^x$ on the same set of axes. 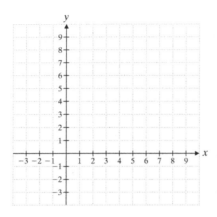	

Helpful Hints

The y-axis is an asymptote for every logarithmic function of the form $f(x) = \log_b x$.

The domain of a logarithmic function $f(x) = \log_b x$ is the set of all positive real numbers, whereas the range of a logarithmic function $f(x) = \log_b x$ is the set of all real numbers.

Concept Check

1. Explain how the domain and range of an exponential function differ from those of a logarithmic function.

Practice

Graph the logarithmic functions.

2. $f(x) = \log_{1/2} x$ 3. $f(x) = \log_e x$ 4. $f(x) = \log_3 x$

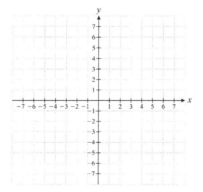 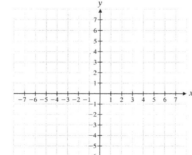 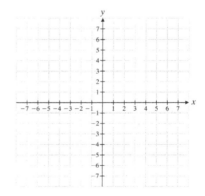

Solving Logarithmic and Exponential Equations
Topic 39.1 Properties of Logarithms

Vocabulary
Logarithm of a Quotient   •   Logarithm of a Product
Logarithm of a Number Raised to a Power

1. The _____ Property states that for any positive real numbers, M and N, and any positive base, $b \neq 1$, $\log_b MN = \log_b M + \log_b N$.

Step-by-Step Video Notes
Watch the Step-by-Step Video lesson and complete the examples below.

Example	Notes
1 & 2. Write each logarithm as a sum of logarithms. $\log_3 AB$ $\log_5 (7 \cdot 11)$	
4. Write the logarithm as a difference of logarithms. $\log_3 \left(\dfrac{29}{7} \right)$ Answer:	
5. Write the expression as a single logarithm. $\log_b 45 - \log_b 9$ Answer:	

Example	Notes
6 & 7. Express each logarithm as a product. $\log_5 b^{10}$ $\log_8 \sqrt{w}$	
8. Write the expression as a single logarithm. $\dfrac{1}{3}\log_b x + 2\log_b w - 3\log_b z$ Answer:	

Helpful Hints

Note that $\left(\log_b M\right)\left(\log_b N\right) \neq \log_b M + \log_b N$ and $\dfrac{\log_b M}{\log_b N} \neq \log_b M - \log_b N$. These are the most common mistakes made when using the properties of logarithms.

For all the properties of logarithms, the base of the logarithm(s) must be the same for the property to apply.

Concept Check

1. Using the properties of exponents, explain why $\log_b \dfrac{M}{N} = \log_b M - \log_b N$.

Practice

Write the expression as a single logarithm.

2. $\log_5 10 + 2\log_5 4.8$

3. $3\log_z \dfrac{1}{2} - \log_z 4$

Write the logarithm as a sum or difference of logarithms.

4. $\log_3\left(x^a r^3 f^n\right)$

5. $\log_c\left(\dfrac{x^3 p^9}{t^5}\right)$

Solving Logarithmic and Exponential Equations
Topic 39.2 Common and Natural Logarithms

Vocabulary

common logarithm   •   natural logarithm   •   base

1. For all real numbers $x > 0$, the _____ of x is $\log_{10} x$, which is often written as $\log x$.

2. For all real numbers $x > 0$, the _____ of x is $\ln x = \log_e x$.

Step-by-Step Video Notes
Watch the Step-by-Step Video lesson and complete the examples below.

Example	Notes
1–3. On a scientific or graphing calculator, approximate the following values. $\log 7.32$ $\log 73.2$ $\log 0.314$	

Example	Notes
4–6. On a scientific or graphing calculator, approximate the following values. $\ln 7.21$ $\ln 72.1$ $\ln 0.0356$	

Helpful Hints
The number e is an irrational number that occurs in many formulas that describe real-world phenomena.

The logarithm, base b, of a positive number, x, is the power (exponent) to which the base b must be raised to produce x and is represented by the expression $\log_b x$.

Concept Check
1. On a scientific or graphing calculator, approximate $\log 2.14$ and $\dfrac{\ln 2.14}{\ln 10}$ to the nearest thousandth. What do you notice?

Practice
On a scientific or graphing calculator, approximate $\log x$ and $\ln x$ for the following values of x. Round to the nearest thousandth as needed.

2. $x = 3.9658$ 4. $x = 100$

3. $x = 1456$ 5. $x = e^{4.286}$

Solving Logarithmic and Exponential Equations
Topic 39.3 Change of Base of Logarithms

Vocabulary
change of base formula   •   common logarithm   •   natural logarithm

1. The _____ states that $\log_b x = \dfrac{\log_a x}{\log_a b}$, where a, b, and $x > 0$, $a \neq 1$, and $b \neq 1$.

Step-by-Step Video Notes
Watch the Step-by-Step Video lesson and complete the examples below.

Example	Notes
1. Evaluate the logarithm using common logarithms. Round to three decimal places. $\log_3 11$ Answer:	
2 & 3. Evaluate each logarithm using common logarithms. Round to three decimal places. $\log_{15} 12$ $\log_7 5.12$	

Example	Notes
4. Evaluate the logarithm using natural logarithms. Round to three decimal places. $\log_4 0.005739$ Answer:	
5 & 6. Evaluate each logarithm using natural logarithms. Round to three decimal places. $\log_{21} 436$ $\log_6 0.315$	

Helpful Hints

It is possible to evaluate $\log_b x$ by determining the power to which b must be raised to produce x, but when x is not a power of b, the change of base formula can be used to approximate the value of $\log_b x$.

To check your answer, y, for the approximation of $\log_b x$, evaluate b^y and check that it is equal to x. Note that when dealing with approximated answers, you should check that b^y is approximately equal to x.

Concept Check

1. Explain why the change of base formula is necessary to compute $\log_2 7$ but not to compute $\log_2 0.125$.

Practice

Evaluate each logarithm using common logarithms. Round to three decimal places.

2. $\log_7 147$

3. $\log_{100} 0.01$

Evaluate each logarithm using natural logarithms. Round to three decimal places.

4. $\log_{14} 8.24$

5. $\log_{4982} 568.2569$

Solving Logarithmic and Exponential Equations
Topic 39.4 Solving Simple Exponential Equations and Applications

Vocabulary
property of exponential equations • compound interest formula
variable compound interest formula • exponential decay formula

1. The _____ states that if $b^x = b^y$, then $x = y$ for $b > 0$ and $b \neq 1$.

Step-by-Step Video Notes
Watch the Step-by-Step Video lesson and complete the examples below.

Example	Notes
1 & 2. Solve the exponential equations. $2^x = 4$ $3^x = 1$	
3 & 4. Solve the exponential equations. $2^x = \dfrac{1}{16}$ $8^{3x-1} = 64$	

Example	Notes
6. If we invest $8000 in a fund that pays 15% annual interest compounded monthly, how much will we have in the account after 6 years?	
7. The decay constant for americium 241 is $k = -0.0016008$. If 10 milligrams (mg) of americium 241 is sealed in a laboratory container today, how much will still be present in 50 years?	

Helpful Hints
To solve an exponential equation, get the same base on both sides of the equation, if possible. Then, set the exponents on each side of the equation equal to each other. If necessary, solve this equation.

If a principle amount, P, is invested at an interest rate, r, that is compounded n times a year, the amount of money, A, accumulated after t years is $A = P\left(1 + \dfrac{r}{n}\right)^{nt}$.

Concept Check
1. Explain how the compound interest formula is a special case of the variable compound interest formula.

Practice
Solve the exponential equations.

2. $5^x = 125$

3. $2^{2x-1} = \dfrac{1}{8}$

4. If a young married couple invests $9000 in a mutual fund that pays 10% interest compounded annually, how much will they have in 5 years?

Solving Logarithmic and Exponential Equations
Topic 39.5 Solving Exponential Equations and Applications

Vocabulary
taking the logarithm of both sides • growth equation • solving exponential equations

1. The _____ Property states that if x and $y > 0$, and $x = y$, then $\log_b x = \log_b y$, where $b > 0$ and $b \neq 1$.

Step-by-Step Video Notes
Watch the Step-by-Step Video lesson and complete the examples below.

Example	Notes
1. Solve the exponential equation. Leave your answer in exact form. $$2^x = 7$$ Take the logarithm of each side of the equation. Solve the equation. Check the solution. Answer:	
3. Solve the exponential equation. Approximate your answer to the nearest ten-thousandth. $$e^{2.5x} = 8.42$$ Take the natural logarithm of each side of the equation. Answer:	

Example	Notes
4. If P dollars is invested in an account that earns interest at 12% compounded annually, the amount available after t years is $A = P(1+0.12)^t$. How many years will it take for $300 dollars in this account to grow to $1500? Round your answer to the nearest whole year.	

Answer:

Helpful Hints
To solve exponential equations, take the logarithm of each side of the equation. Then, solve the resulting equation using the properties of logarithms and check the solution.

The growth equation for populations that grow continuously is $A = A_0 e^{rt}$, where A is the amount at time, t, A_0 is the original amount, r is the rate at which things are growing in a unit of time, and t is the total number of time in years.

Concept Check
1. Solve $10^x = 1$ by taking the common logarithm of both sides. Then, solve it by converting it to a logarithmic equation. Do you get the same answer? Explain.

Practice
Solve the exponential equation. Approximate your answer to the nearest thousandth.

2. $e^{2.5x} = 19$

3. $4^{4x+1} = 28$

4. $110^x = 14.58$

5. If the population of mosquitoes in a small town is nine billion and it continues to grow at a rate of 3% per year, how many years will it take for the population to increase to fifteen billion mosquitoes? Round your answer to the nearest whole year.

Solving Logarithmic and Exponential Equations
Topic 39.6 Solving Simple Logarithmic Equations and Applications

Vocabulary

logarithmic equation   •   exponential equation   •   logarithm

1. To solve a(n) _____, convert it to a(n) _____ and solve.

Step-by-Step Video Notes

Watch the Step-by-Step Video lesson and complete the examples below.

Example	Notes
1. Solve the logarithmic equation $\log_2 x = 4$. The exponential equation is $\Box^{\Box} = x$. $x = \Box$ Check the solution. Answer:	
2–4. Solve the logarithmic equations. $3 = \log_3 x$ $\log_{10} x = -3$ $\log_{25} x = \dfrac{1}{2}$	

Example	Notes
5 & 6. Solve the logarithmic equations. $\log_7 343 = y$ $\log_8\left(\dfrac{1}{64}\right) = y$	

7 & 8. Solve the logarithmic equations.

$2 = \log_b 121$

$\log_b 81 = 4$

Helpful Hints

To solve an exponential equation, remember to first get the same bases on both sides of the equation. Then set the exponents equal to each other.

To solve an exponential equation for the base, try to get the exponents on both sides of the equation equal, and then set the bases equal to each other.

Concept Check

1. While solving the equation $\log_b 4 = 2$, Robert got the answer $b = 16$. Explain his error.

Practice

Solve the logarithmic equations.

2. $\log_3 x = 4$

3. $\log_8 x = \dfrac{1}{3}$

4. $\log_b 128 = 7$

5. $\log_5\left(\dfrac{1}{125}\right) = y$

Solving Logarithmic and Exponential Equations
Topic 39.7 Solving Logarithmic Equations and Applications

Vocabulary

Logarithmic Property of Equality   •   logarithmic equation

1. The _____ states that if $\log_b x = \log_b y$, then $x = y$, where $b > 0$ and $b \neq 1$, and x and y are positive real numbers.

Step-by-Step Video Notes
Watch the Step-by-Step Video lesson and complete the examples below.

Example	Notes
1. Solve $\log 5 = 2 - \log(x+3)$ for x. Isolate the logarithm. Convert the equation to an exponential equation. Solve the equation. Check the solution(s). Answer:	
3. Solve for x. $\log(x+6) + \log(x+2) = \log(x+20)$ Answer:	

Example	Notes
4. The magnitude of an earthquake is measured by the formula $R = \log\left(\dfrac{I}{I_0}\right)$, where I is the intensity of the earthquake and I_0 is the minimum measurable intensity. The 1964 earthquake in Anchorage, Alaska, had a magnitude of 8.4. The 1906 earthquake in Taiwan had a magnitude of 7.1. How many times more intense was the Anchorage earthquake than the Taiwan earthquake?	

Answer:

Helpful Hints
To solve logarithmic equations, if there is more than one logarithmic term, use the properties of logarithms to rewrite them as a single logarithm. Then, get one logarithmic term on one side of the equation and one numerical value on the other. Finally, convert the equation to an exponential equation and solve. Be sure to check any solutions.

To solve logarithmic equations with only logarithmic terms, use the properties of logarithms to write each side as a single logarithm. Then, use the logarithmic property of equality and solve the resulting equation. Be sure to check any solutions.

Concept Check
1. When solving the equation $\log_6 (x - 3) + \log_6 (x + 2) = 1$ for x, Paul found that $x = 4$ and $x = -3$. Explain Paul's error.

Practice
Solve for x.

2. $\log_4 (x + 3) + \log_4 (x - 3) = 2$

4. How many times more intense is a magnitude 5.5 earthquake than a magnitude 5.1 earthquake?

3. $\log_7 (x + 6) - \log_7 (x + 2) = \log_7 (x + 3)$

U.S. and Metric Measurement
Topic A.1 U.S. Length

Vocabulary

conversion   •   unit fraction   •   length   •   mixed unit

1. A _____ has a value of 1. Its numerator and denominator express the same value in different ways.

Step-by-Step Video Notes
Watch the Step-by-Step Video lesson and complete the examples below.

Example	Notes
1. Convert 24 yards to feet using unit fractions.	

Table 2 U.S. Length Conversions
12 in. = 1ft
36 in. = 1 yd
3 ft = 1 yd
5280 ft = 1 mi
1760 yd = 1 mi

There are $\square$ ft in 1 yd. Using

$$\frac{\text{unit of measurement converting to}}{\text{original unit of measurement}}, \text{ the}$$

numerator is in ft, and the denominator is in yd.

$$24 \text{ yards} \cdot \frac{\square \text{ feet}}{1 \text{ yard}} = \square \text{ feet}$$

2. Convert 15,840 ft to mi using unit fractions.

There are $\boxed{}$ feet in 1 mile.

$$15,840 \text{ ft} \cdot \frac{1 \text{ mi}}{\boxed{} \text{ ft}} = \square \text{ mi}$$

Example	Notes
4. Convert the following using unit fractions. 90 in. to yd $90 \text{ in.} \cdot \dfrac{\boxed{}\underline{}}{\boxed{}\underline{}} = \boxed{} \text{ yd}$ Answer:	
5. Gage is 54 inches tall. Convert this to feet and inches. $54 \text{ in.} \cdot \dfrac{\boxed{}\underline{}}{\boxed{}\underline{}} = \boxed{} \text{ r } \boxed{}$ The quotient is the number of feet, and the remainder is in inches. Answer:	

Helpful Hints

Use unit fractions such as $\dfrac{12 \text{ in.}}{1 \text{ ft}}$, $\dfrac{36 \text{ in.}}{1 \text{ yd}}$, $\dfrac{3 \text{ ft}}{1 \text{ yd}}$, $\dfrac{5280 \text{ ft}}{1 \text{ mi}}$, and $\dfrac{1760 \text{ yd}}{1 \text{ mi}}$ to convert units.

The fraction $\dfrac{\text{unit of measurement converting to}}{\text{original unit of measurement}}$ can be helpful to use when converting units.

Concept Check
1. Why is a unit fraction equal to 1 if the numerator and denominator have different numbers?

Practice

Convert the following using unit fractions.
2. 468 in. to yd

3. 3.1 miles to yards

Convert the heights to feet and inches.
4. Ben is 68 inches tall.

5. Gomez is 76 inches tall.

Name: _____   Date: _____

Instructor: _____   Section: _____

U.S. and Metric Measurement
Topic A.2 U.S. Weight and Capacity

Vocabulary

weight   •   mass   •   capacity   •   pound   •   gallon

1. _____ is related to the gravitational pull on an object.

2. _____ is the amount of space inside a three-dimensional figure.

Step-by-Step Video Notes
Watch the Step-by-Step Video lesson and complete the examples below.

Example	Notes
1. Convert 7.5 tons to pounds.	

U.S. Weight Conversions

16 oz = 1 lb
2000 lbs = 1 ton

There are ☐ pounds in 1 ton.

Using $\dfrac{\text{unit of measurement converting to}}{\text{original unit of measurement}}$, the

numerator is in pounds, and the denominator is in tons.

$7.5 \text{ tons} \cdot \dfrac{\boxed{} \text{ pounds}}{1 \text{ ton}} = \boxed{} \text{ pounds}$

Answer:

2. Convert 64 oz to lb.

There are ☐ oz in 1 lb.

$64 \text{ oz} \cdot \dfrac{1}{\boxed{} \text{ oz}} = \boxed{}\;\boxed{}$

Answer:

Example	Notes

4. Convert 26 quarts to gallons.

Table 4 U.S. Capacity Conversions
8 fluid ounces = 1 cup
2 cups = 1 pint
16 fluid ounces = 1 pint
2 pints = 1 quart
4 quarts = 1 gallon

Give your answer in decimal or fraction form.

Answer:

5. Lashonda buys a bottle of ketchup that contains 44 fl oz. Convert this to pints and fluid ounces.

The quotient is the number of pints, and the remainder is in fluid ounces.

Answer:

Helpful Hints

Use unit fractions such as $\dfrac{16 \text{ oz}}{1 \text{ lb}}$, $\dfrac{2000 \text{ lb}}{1 \text{ ton}}$, $\dfrac{8 \text{ fl oz}}{1 \text{ cup}}$, $\dfrac{16 \text{ fl oz}}{1 \text{ pint}}$, $\dfrac{1 \text{ quart}}{2 \text{ pints}}$, and $\dfrac{4 \text{ quarts}}{1 \text{ gallon}}$ to convert units. Note that weight and capacity are often given in mixed units, such as pounds and ounces, pints and fluid ounces, etc.

Concept Check

1. Why use pounds and ounces for the weight of a newborn baby, rather than a decimal number, as is often used with weights measured in tons?

Practice

Convert the following to pounds.

2. 96 ounces

3. 3.7 tons

Convert the following to gallons.

4. 64 fl oz

5. 36 quarts

U.S. and Metric Measurement
Topic A.3 Metric Length

Vocabulary

metric prefixes   •   meter   •   milli-   •   centi-   •   kilo-
deka-   •   hecto-   •   deci-

1. The metric prefix _____ means 1000.

Step-by-Step Video Notes
Watch the Step-by-Step Video lesson and complete the examples below.

Example	Notes

1. Convert 400 centimeters to meters.

Table 1 Metric Units of Length	
Conversion	**Unit Fraction**
1000 millimeters (mm) = 1 meter (m)	$\dfrac{1 \text{ meter}}{1000 \text{ millimeters}}$ or $\dfrac{1000 \text{ millimeters}}{1 \text{ meter}}$
100 centimeters (cm) = 1 meter (m)	$\dfrac{1 \text{ meter}}{100 \text{ centimeters}}$ or $\dfrac{100 \text{ centimeters}}{1 \text{ meter}}$
10 decimeters (dm) = 1 meter (m)	$\dfrac{1 \text{ meter}}{10 \text{ decimeters}}$ or $\dfrac{10 \text{ decimeters}}{1 \text{ meter}}$
1 meter (m) is the basic unit of length	
10 meters (m) = 1 dekameter (dam)	$\dfrac{10 \text{ meter}}{1 \text{ dekameters}}$ or $\dfrac{1 \text{ dekameters}}{10 \text{ meter}}$
100 meters (m) = 1 hectometer (hm)	$\dfrac{100 \text{ meter}}{1 \text{ hectometer}}$ or $\dfrac{1 \text{ hectometer}}{100 \text{ meter}}$
1000 meters (m) = 1 kilometer (km)	$\dfrac{1000 \text{ meter}}{1 \text{ kilometer}}$ or $\dfrac{1 \text{ kilometer}}{1000 \text{ meter}}$

There are ☐ centimeters in 1 meter.

$$400 \text{ centimeters} \cdot \frac{1 \text{ meter}}{100 \text{ centimeters}} = \boxed{} \text{ meters}$$

Answer:

Example	Notes
3. Convert 96.3 km to m. List the prefixes. <u>km</u> hm dam <u>m</u> dm cm mm To convert from km to m, move the decimal point ☐ places to the _____. Answer:	
4. Convert 150 dam to cm. 150 dam = ☐☐☐,☐☐☐ cm Answer:	

Helpful Hints

The metric system is based on powers of 10. Converting units can be done by moving the decimal point. The mnemonic "Kangaroos hopping down mountains drinking chocolate milk" can help you remember the metric prefixes in order from largest to smallest, kilometers, hectometers, decameters, meter, decimeter, centimeter, millimeter.

List the prefixes like this km hm dam m dm cm mm is a visual way to tell which way to move the decimal point when converting metric units of length. Start with the original unit and move to the new unit. Move the decimal point accordingly, the same number of spaces and in the same direction, adding zeros as necessary.

Concept Check

1. When converting from meters to centimeters, how many places and in what direction should you move the decimal point?

Practice

Convert the following to meters.

2. 87 cm

3. 6.2 km

Convert the following to centimeters.

4. 25 m

5. 44 mm

U.S. and Metric Measurement
Topic A.4 Metric Mass and Capacity

Vocabulary
gram • mass • kilogram • milligram • dekagram
liter • milliliter • capacity • deciliter • kiloliter

1. The basic unit of mass in the metric system is the _____.

2. A _____ is slightly more than two pounds.

Step-by-Step Video Notes
Watch the Step-by-Step Video lesson and complete the examples below.

Example	Notes
1. Convert 32 centigrams to grams. There are ☐ centigrams in 1 gram. Using $\dfrac{\text{unit of measurement converting to}}{\text{original unit of measurement}}$, the numerator is in grams, and the denominator is in centigrams. $32\ \text{cg} \cdot \dfrac{1\ \text{g}}{\boxed{}\ \text{cg}} = \boxed{}\ \text{g}$ Answer:	
3. Convert 216 kg to cg. List the prefixes. <u>kg hg dag g dg cg mg</u> To convert from kg to cg, move the decimal point ☐ places to the _____. Add zero(s) to the end of the decimal to move the decimal point the correct number of places. Answer:	

Example	Notes
4. Convert 900 mL to L. List the prefixes. kL hL daL <u>L dL cL mL</u> To convert from mL to L, move the decimal point ☐ places to the _____. Answer:	
5. Convert 83.2 L to cL. Answer:	

Helpful Hints
Regardless of the type of measure (length, mass, or capacity), the metric prefixes always have the same meaning and relationship to the basic unit.

With mass and capacity, the prefixes kilo- and milli- are most often used.

Concept Check
1. Which metric unit is closest to a quart in U.S. measurement? Is a gallon more or less than 4 liters?

Practice

Convert the following to grams.
2. 96 mg

3. 5.8 kilograms

Convert the following to liters.
4. 500 mL

5. 4.4 kL

Name: _____   Date: _____

Instructor: _____   Section: _____

U.S. and Metric Measurement
Topic A.5 Converting Between U.S. and Metric Units

Vocabulary
meter   •   gallon   •   pound   •   approximately   •   exactly

1. The symbol ≈ means _____.

Step-by-Step Video Notes
Watch the Step-by-Step Video lesson and complete the examples below.

Example	Notes
1. Convert 44.02 miles to kilometers.	

Table 1 U.S. to Metric (Length)	
Conversion	**Unit Fraction**
1 inch (in.) = 2.54 centimeters (cm)	$\dfrac{2.54 \text{ cm}}{1 \text{ in.}}$
1 foot (ft) = 0.30 meter (m)	$\dfrac{0.30 \text{ m}}{1 \text{ ft}}$
1 yard (yd) = 0.91 meter (m)	$\dfrac{0.91 \text{ m}}{1 \text{ yd}}$
1 mile (mi) = 1.61 kilometers (km)	$\dfrac{1.61 \text{ km}}{1 \text{ mi}}$

Table 2 Metric to U.S. (Length)	
Conversion	**Unit Fraction**
1 meter (m) = 39.37 inches (in.)	$\dfrac{39.37 \text{ in.}}{1 \text{ m}}$
1 meter (m) = 1.09 yards (yd)	$\dfrac{1.09 \text{ yd}}{1 \text{ m}}$
1 meter (m) = 3.28 feet (ft)	$\dfrac{3.28 \text{ ft}}{1 \text{ m}}$
1 kilometer (km) = .62 mile (mi)	$\dfrac{62 \text{ mi}}{1 \text{ km}}$

There are about ☐ kilometers in 1 mile.

Convert using the unit fraction.

$$44.02 \text{ mi} \cdot \frac{1.61 \text{ km}}{1 \text{ mi}} = \boxed{} \text{ km}$$

Answer:

Example	Notes

3. Convert 5.2 kg to lb. Use the unit fraction

Table 3 U.S. to Metric (Weight/Mass)	
Conversion	**Unit Fraction**
1 ounce (oz) = 28.35 grams (g)	$\frac{28.35 \text{ g}}{1 \text{ oz}}$
1 pound (lb) = 0.45 kilogram (kg)	$\frac{0.45 \text{ kg}}{1 \text{ lb}}$

Table 4 Metric to U.S. (Weight/Mass)	
Conversion	**Unit Fraction**
1 kilogram (kg) = 2.20 pounds (lb)	$\frac{2.20 \text{ lb}}{1 \text{ kg}}$
1 gram (g) = 0.035 ounce (oz)	$\frac{0.035 \text{ oz}}{1 \text{ g}}$

Answer:

6. Convert 3 qt to L. Use the unit fraction to convert.

Table 5 U.S. to Metric (Capacity)	
Conversion	**Unit Fraction**
1 quart (qt) = 0.95 liter (L)	$\frac{0.95 \text{ L}}{1 \text{ qt}}$
1 gallon (gal) = 3.79 liters (L)	$\frac{3.79 \text{ L}}{1 \text{ gal}}$

Table 6 Metric to U.S. (Capacity)	
Conversion	**Unit Fraction**
1 liter (L) = 1.06 quarts (qt)	$\frac{1.06 \text{ qt}}{1 \text{ L}}$
1 liter (L) = 0.26 gallon (gal)	$\frac{0.26 \text{ gal}}{1 \text{ L}}$

Answer:

Helpful Hints

Almost all unit fractions used to convert between U.S. and metric units are approximate. Not all tables provide every conversion fact. You may need to change to units provided in the table, then use another conversion fact you know to complete the conversion.

Concept Check

1. When measuring a length, will there be more units if you measure in yards or in meters?

Practice

Convert the following to meters.

2. 100 yards

3. 6.2 miles

Convert the following to quarts.

4. 18 L

5. 947 mL

U.S. and Metric Measurement
Topic A.6 Time and Temperature

Vocabulary
conversion   •   unit fraction   •   denominator   •   equivalent fraction

1. A(n) _____ has a value of 1. Its numerator and denominator express the same value in different ways.

Step-by-Step Video Notes
Watch the Step-by-Step Video lesson and complete the examples below.

Example	Notes
1. Convert 195 minutes to hours. When using unit fractions to make conversions, use the following fraction. $$\dfrac{\text{unit of measurement converting to}}{\text{original unit of measurement}}$$ $195 \text{ minutes} \times \dfrac{1 \text{ hour}}{\boxed{} \text{ minutes}}$ Answer:	
2. Convert 4.4 hours to seconds. Answer:	
3. Convert 36° Fahrenheit to Celsius. Round to the nearest tenth. Substitute 36 for F in the formula. $C = \dfrac{5 \times \boxed{} - 160}{9}$ Answer:	

Example	Notes
5. Marques, who lives in the United States, purchased a heat pump that was made in Belgium. The manual states that the heat pump can operate efficiently up to a temperature of 39° Celsius. To the nearest degree, what is this temperature in Fahrenheit? Substitute 39 into the formula for converting Celsius to Fahrenheit. $F = 1.8\left(\boxed{}\right) + 32$ Answer:	

Helpful Hints

The formula for converting Fahrenheit to Celsius is $C = \dfrac{5 \times F - 160}{9}$. The formula for converting Celsius to Fahrenheit is $F = 1.8C + 32$.

It may be necessary to use more than one unit fraction in a conversion.

Concept Check

1. What unit fraction(s) would you use to convert days to minutes?

Practice

2. Convert 145 minutes to hours.

3. Convert 2 years to hours.

4. Convert 100° C to Fahrenheit.

5. Seamus is spending next week in the United States for a school trip. He reads the average temperature will be 95° F. To the nearest degree, what is this temperature in Celsius?

Name: _____   Date: _____

Instructor: _____   Section: _____

More on Functions
Topic B.1 Using a Graphing Calculator

Vocabulary
viewing window   •   standard window   •   square window   •   dot mode
connected mode   •   intersection   •   zoom   •   trace   •   caret

1. The _____ has the settings $X_{min} = -10$, $X_{max} = 10$, $X_{scl} = 1$, $Y_{min} = -10$, $Y_{max} = 10$, and $Y_{scl} = 1$.

2. In addition to the "_____" feature, the "zoom" and "trace" features can also be used to find the intersection of two graphs.

3. When your calculator is in _____ mode, remember that the vertical lines are not part of the graph; instead, they represent vertical asymptotes.

Step-by-Step Video Notes
Watch the Step-by-Step Video lesson and complete the examples below.

Example	Notes
2. Graph $3x + 2y = 6$ on a graphing calculator. Solve the equation for y. Input the equation. $y_1 = \boxed{(}\ \boxed{}\ \boxed{-}\ \boxed{}\ \boxed{x}\ \boxed{)}\ \boxed{\div}\ \boxed{}$ $\boxed{\text{GRAPH}}$ 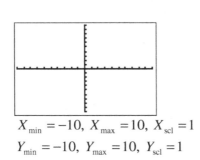 $X_{min} = -10,\ X_{max} = 10,\ X_{scl} = 1$ $Y_{min} = -10,\ Y_{max} = 10,\ Y_{scl} = 1$	

Example	Notes

4. Solve $x = y^2 - 3$ for y and graph.

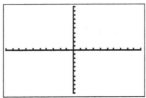

$X_{min} = -10,\ X_{max} = 10,\ X_{scl} = 1$
$Y_{min} = -10,\ Y_{max} = 10,\ Y_{scl} = 1$

---

5. Graph $y = x^2 + 3x - 10$ on a graphing calculator and find the vertex.

Open the graphing screen and input the equation.

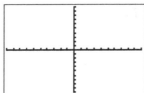

$X_{min} = -10,\ X_{max} = 10,\ X_{scl} = 1$
$Y_{min} = -10,\ Y_{max} = 10,\ Y_{scl} = 1$

Adjust the window settings because the entire graph is not in the window.

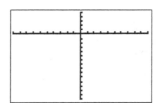

$X_{min} = -10,\ X_{max} = 10,\ X_{scl} = 1$
$Y_{min} = -15,\ Y_{max} = 5,\ Y_{scl} = 1$

Calculate the minimum.

Answer:

Example	Notes

Example

7. Graph the function $f(x) = \dfrac{9x-12}{(x+3)(x-4)}$ in connected mode and identify the domain and y-intercept.

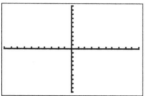

$X_{min} = -10,\ X_{max} = 10,\ X_{scl} = 1$
$Y_{min} = -10,\ Y_{max} = 10,\ Y_{scl} = 1$

Answer:

9 & 10. In each of the following, use your graphing calculator to determine if y_1 is equivalent to y_2.

$y_1 = \dfrac{2x+1}{x+3} - \dfrac{5x}{5x+15}$ and $y_2 = \dfrac{x+1}{x+3}$

$y_1 = \dfrac{3x}{2x-3} + \dfrac{3x+6}{2x^2+x-6}$ and $y_2 = \dfrac{3x+3}{2x+3}$

13. Solve $\sqrt{4-x} = x-2$ by finding the point of intersection of $y_1 = \sqrt{4-x}$ and $y_2 = x-2$.

Graph y_1 and y_2. Use the zoom feature to zoom in to the point of intersection. Use the trace feature to find the point of intersection.

Answer:

Helpful Hints
One limitation that is common to almost all graphing calculators is that they only understand equations that are solved for *y*.

On a graphing calculator, asymptotes are often displayed as vertical lines (in connected mode) or not displayed (in dot mode).

When graphing rational functions, be sure to use parentheses around the numerator and denominator of each rational expression.

When graphing radical functions, be sure to use parentheses around the radicand.

Concept Check
1. Would you use a graphing calculator to find the *y*-interecept of the function
    $g(x) = -2x + 8$? Explain.

Practice
2. Graph $4x - 3y = 12$ on a graphing calculator.

4. Solve $\sqrt{2-x} = 2x + 6$ by finding the point of intersection of $y_1 = \sqrt{2-x}$ and
    $y_2 = 2x + 6$.

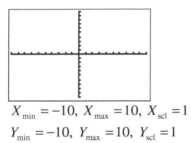

$X_{min} = -10, \ X_{max} = 10, \ X_{scl} = 1$
$Y_{min} = -10, \ Y_{max} = 10, \ Y_{scl} = 1$

3. Solve $x = y^2 + 2$ for *y* and graph.

5. Graph the function $f(x) = \dfrac{x+1}{(x-2)(x+5)}$

    in dot mode and identify the domain and range.

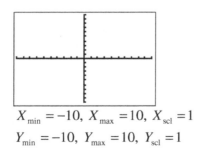

$X_{min} = -10, \ X_{max} = 10, \ X_{scl} = 1$
$Y_{min} = -10, \ Y_{max} = 10, \ Y_{scl} = 1$

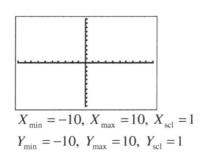

$X_{min} = -10, \ X_{max} = 10, \ X_{scl} = 1$
$Y_{min} = -10, \ Y_{max} = 10, \ Y_{scl} = 1$

More on Functions
Topic B.2 Algebra of Functions

Vocabulary
function notation   •   algebraic operations on functions   •   domain

1. Just as they can be performed on real numbers, _____ can be performed, which consist of the sum, difference, product, and quotient of two functions.

Step-by-Step Video Notes
Watch the Step-by-Step Video lesson and complete the examples below.

Example	Notes
1 & 2. For $f(x) = 2x$ and $g(x) = x+1$, perform the indicated operations. $(f+g)(x) = f(x) + \boxed{}$ $\qquad\qquad = 2x + \boxed{}$ $(f-g)(x) = \boxed{} - \boxed{}$	
3 & 4. For $f(x) = 4x+3$ and $g(x) = x+12$, perform the indicated operations. $(f \cdot g)(-3)$ $\left(\dfrac{f}{g}\right)(-12)$	

Example	Notes
5. For $p(x) = 3x + 9$ and $q(x) = 2x - 11$, perform the indicated operations. $(p+q)(2)$	
8. For $f(x) = 7x^2 + 30x + 8$ and $g(x) = x + 4$, perform the indicated operation. $\left(\dfrac{f}{g}\right)(x)$	

Helpful Hints

Do not confuse the function notation $f(x)$ with notation for multiplication; $f(x)$ does not mean "f multiplied by x."

When performing operations on functions, the domains of the functions intersect. When dividing two functions, be sure to exclude from the domain the values that make the function in the denominator equal to zero.

Concept Check

1. For $f(x) = 8x + 5$ and $g(x) = x^2 - 1$, find the domain of $\left(\dfrac{f}{g}\right)(x)$ without actually performing the division. Explain.

Practice

For $f(x) = 7x + 10$ and $g(x) = 2x^2 - 5x - 3$, perform the indicated operations.

2. $(f+g)(x)$

4. $(f \cdot g)(x)$

3. $(f-g)(-2)$

5. $\left(\dfrac{f}{g}\right)(2)$

Name: _____   Date: _____

Instructor: _____   Section: _____

More on Functions
Topic B.3 Transformations of Functions

Vocabulary
transformation • vertical shift • horizontal shift • reflection • absolute value

1. A(n) _____ in a graph is when the graph has been shifted either upward or downward from the original graph.

2. The graph of $g(x)$ is a(n) _____ of the graph of $f(x)$ if $g(x) = -f(x)$.

Step-by-Step Video Notes
Watch the Step-by-Step Video lesson and complete the examples below.

Example	**Notes**
1. Graph $h(x) = x^2 + 2$ by applying a vertical shift to the graph of $f(x) = x^2$. 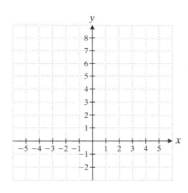	
2. Graph $p(x) = \|x - 3\|$ by applying a horizontal shift to the graph of $f(x) = \|x\|$. 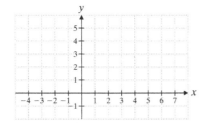	

Example	Notes

4. Graph $g(x) = -x^2$ by transforming the graph of $f(x) = x^2$.

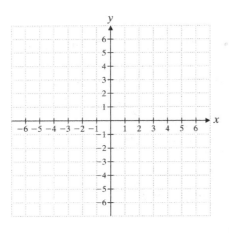

Helpful Hints

If a function is described by $f(x \pm h) \pm k$, then it is both a vertical and a horizontal shift on the function $f(x)$.

Concept Check

1. Pete claims that the graph of $g(x) = x + 5$ is a vertical shift of $f(x) = x$, but April claims that $g(x)$ is a horizontal shift of $f(x)$. Who is correct? Explain.

Practice

Graph each of the following by transforming the graph of $f(x) = x^3$.

2. $g(x) = (x+2)^3$

3. $h(x) = (x-1)^3 + 3$

4. $p(x) = -x^3$

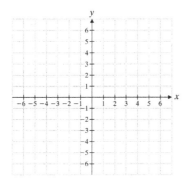 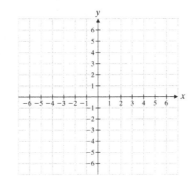 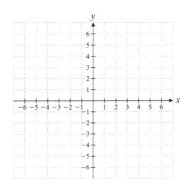

More on Functions
Topic B.4 Piecewise Functions

Vocabulary
independent variable   •   piecewise function   •   absolute value

1. A(n) _____ is a function whose definition changes depending on the value of the independent variable.

Step-by-Step Video Notes
Watch the Step-by-Step Video lesson and complete the examples below.

Example	Notes
1. If $f(x) = \begin{cases} 2x & \text{if} & x < 2 \\ x^2 & \text{if} & x \geq 2 \end{cases}$, find the following. $f(3)$ Determine which domain the value fits into, then substitute 3 for x in the corresponding expression and simplify. Answer:	
2. If $f(x) = \begin{cases} 2x+3 & \text{if } x \leq 0 \\ -x-1 & \text{if } x > 0 \end{cases}$, find the following. $f(2)$ $f(-6)$ $f(0)$	

Example	Notes
3. Susan's weekly salary (in dollars) is given by $f(x) = \begin{cases} 10x & \text{if } 0 \le x \le 40 \\ 20x - 400 & \text{if } x > 40 \end{cases}$, where x is the number of hours worked per week. Find how much money Susan will make if she works the following number of hours in a week. 30 hours 60 hours	
4. Write the absolute value function $f(x) = |x|$ as a piecewise function.	

Helpful Hints
Each x value of a function must have one and only one y value. Do not substitute the value of x into more than one expression when evaluating a function.

Concept Check

1. Is $f(x) = \begin{cases} x^2 & \text{if } x < 0 \\ 4x^2 - 3x^2 & \text{if } x \ge 0 \end{cases}$ a piecewise function? Explain.

Practice
Find the value of $f(2)$ for each of the following piecewise functions.

2. $f(x) = \begin{cases} 5x - 7 & \text{if } x \le 2 \\ -x + 4 & \text{if } x > 2 \end{cases}$ 3. $f(x) = \begin{cases} x^2 + 6 & \text{if } x \le 0 \\ -x^2 - 1 & \text{if } x > 0 \end{cases}$ 4. $f(x) = \begin{cases} 0.5x & \text{if } x > 0 \\ x^{16} & \text{if } x \le 0 \end{cases}$

More on Functions
Topic B.5 Graphing Piecewise Functions

Vocabulary

piecewise function   •   domain   •   horizontal line

1. A _____ is a function whose definition changes depending on the value of the independent variable.

Step-by-Step Video Notes
Watch the Step-by-Step Video lesson and complete the examples below.

Example	Notes

2. Graph $f(x) = \begin{cases} 2x+3 & \text{if} \quad x \le 0 \\ -x-1 & \text{if} \quad x > 0 \end{cases}$.

Make a table of values for each piece.

x	y
0	
-1	
-2	

x	y
0	
1	
2	
3	

Graph each piece.

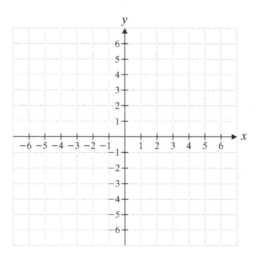

Example | **Notes**

4. Graph $f(x) = \begin{cases} -x & \text{if} & x \le -1 \\ 2x & \text{if} & -1 < x \le 2 \\ 4 & \text{if} & x > 2 \end{cases}$

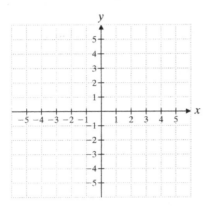

Helpful Hints

To evaluate a piecewise function at a certain value, determine which domain the value fits into, and then substitute the value of the variable into that correct expression and simplify.

To graph a piecewise function, graph each piece separately on the same set of axes, paying special attention to the starting points and ending points for each.

Concept Check

1. You are asked to graph $f(x) = \begin{cases} 5x+1 & \text{if} & x \le -1 \\ 2x-3 & \text{if} & x > -2 \end{cases}$. Is this a piecewise function? Why or why not?

Practice

Graph each piecewise function.

2. $f(x) = \begin{cases} 2 & \text{if} & x \le 2 \\ x & \text{if} & x > 2 \end{cases}$

3. $f(x) = \begin{cases} x^2 & \text{if} & x < 0 \\ -x^2 & \text{if} & x \ge 0 \end{cases}$

4. $f(x) = \begin{cases} -3 & \text{if} & x < -4 \\ |x| & \text{if} & -4 \le x \le 2 \\ -x+4 & \text{if} & x > 2 \end{cases}$

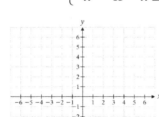 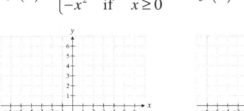 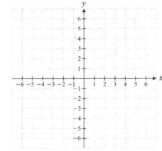

More on Systems
Topic C.1 Systems of Linear Inequalities

Vocabulary
system of linear inequalities   •   test point   •   point of intersection

1. Two or more linear inequalities graphed on the same set of axes is called a

   _____.

Step-by-Step Video Notes
Watch the Step-by-Step Video lesson and complete the examples below.

Example	**Notes**
2. Graph the solution to the system of linear inequalities. $y < -\dfrac{4}{3}x + 3$ $y \ge \dfrac{1}{2}x - 2$ 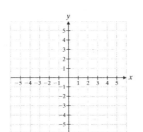	
3. Graph the solution to the system of linear inequalities. $y \ge 2x + 3$ $4x - 2y > 4$ 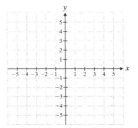	

Example	Notes
4. Graph the solution to the system of linear inequalities. Find and label all the points of intersection. $x + y \leq 3$ $x - y \leq 1$ $x \geq -1$ 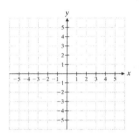	

Helpful Hints
The solution to a system of linear inequalities is the intersection of the solution sets of the individual inequalities.

If the graph of a system of linear inequalities is parallel lines with the shading outside the lines, there is no solution to the system.

Concept Check
1. Can a system of two linear inequalities have no solution if it has two perpendicular lines?

Practice
Graph the solution to the given system of inequalities.

2. $y \geq 0.5x$
   $x \geq 2y + 3$

3. $y \geq 3$
   $x < -1$

4. $y - x > 2$
   $y < 2$
   $y \geq -2x - 4$

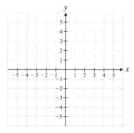

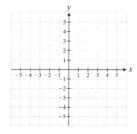

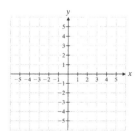

More on Systems
Topic C.2 Systems of Non-Linear Equations

Vocabulary
solution   •   system of non-linear equations   •   substitution   •   elimination method

1. A(n) _____ is a system of equations where at least one of the equations is not linear.

Step-by-Step Video Notes
Watch the Step-by-Step Video lesson and complete the examples below.

Example	Notes
1. Solve by graphing. $y = x^2$ $y = x + 6$ 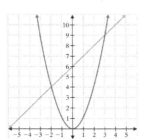 Solution:	
2. Solve using the substitution method. $y = x^2$ $y = 2x + 15$ $y = 2x + 15$ $\boxed{} = 2x + 15$ ← equation in x Solution:	

Example	Notes
3. Solve using the elimination method. $2x^2 - 5y^2 = -2$ $3x^2 + 2y^2 = 35$ Solution:	
4. Solve by any method. $x^2 + y^2 = 4$ $y = x^2 - 2$ Solution:	

Helpful Hints

Solving a system of non-linear equations by substitution is usually easiest if one of the equations is linear, or if the equation is already solved for one variable.

In order to solve a system by elimination, the variables in both equations must be the same, and the terms must be like terms.

Concept Check

1. Refer to Example 1, if the graphing method were not used, would you solve this system with the substitution method or the elimination method? Explain.

Practice

2. Solve by any method.

$y = -x^2$
$y = -x - 2$

3. Solve by any method.

$y = -2x$
$y = 8x^2 - 15$

4. Solve by any method.

$3x^2 - 7y^2 = -25$
$5x^2 + 3y^2 = 17$

More on Systems
Topic C.3 Matrices and Determinants

Vocabulary
matrix • dimensions of a matrix • determinant • value of a 2×2 determinant
value of a 3×3 determinant • minor of an element of a 3×3 determinant • array of signs

1. A(n) _____ is a rectangular array of numbers that is arranged in rows and columns.

2. The _____ $\begin{vmatrix} a & c \\ b & d \end{vmatrix}$ is $ad - bc$.

Step-by-Step Video Notes
Watch the Step-by-Step Video lesson and complete the examples below.

Example	Notes
1. Write the coefficients of the variables in the system of equations as a matrix. $\begin{array}{r} 5y = 10 \\ 2x - 6y = 17 \end{array}$ $\begin{bmatrix} \square & \square \\ \square & \square \end{bmatrix}$	
3. Find the determinant of the matrix. $\begin{bmatrix} 0 & -3 \\ -2 & 6 \end{bmatrix}$ Answer:	
4. Evaluate the determinant. $\begin{vmatrix} 4 & 1 & 2 \\ 3 & -1 & 0 \\ 1 & 2 & 3 \end{vmatrix}$ Answer:	

Example	Notes
7. Evaluate the determinant by expanding it by minors of elements in the first column. $$\begin{vmatrix} 2 & 3 & 6 \\ 4 & -2 & 0 \\ 1 & -5 & -3 \end{vmatrix}$$ Answer:	

Helpful Hints

To evaluate a 3×3 determinant, use expansion by minors of elements in the first column.

The minor of an element (number or variable) of a 3×3 determinant is the 2×2 determinant that remains after we delete the row and column in which the element appears.

Concept Check

1. What does the array of signs help to determine?

Practice

Evaluate the determinants.

2. $\begin{vmatrix} 4 & 0 \\ 10 & -1 \end{vmatrix}$

3. $\begin{vmatrix} 3 & 4 \\ -1 & -2 \end{vmatrix}$

Evaluate the determinant by expanding by minors.

4. $\begin{vmatrix} 1 & 3 & 4 \\ -2 & 0 & 0 \\ -4 & 5 & -1 \end{vmatrix}$

More on Systems
Topic C.4 Solving Systems of Linear Equations Using Matrices

Vocabulary
augmented matrix   •   matrix row operation

1. A matrix that is derived from a system of linear equations is called the _____ of the system.

Step-by-Step Video Notes
Watch the Step-by-Step Video lesson and complete the examples below.

Example	**Notes**
1. Solve the system of equations using a matrix. $4x - 3y = -13$ $x + 2y = 5$ Write the augmented matrix and perform row operations to solve the system of equations. $\begin{bmatrix} \Box & \Box & \Box \\ \Box & \Box & \Box \end{bmatrix} R_1 \leftrightarrow R_2 \begin{bmatrix} \Box & \Box & \Box \\ \Box & \Box & \Box \end{bmatrix}$ $\begin{bmatrix} \Box & \Box & \Box \\ \Box & \Box & \Box \end{bmatrix} -4R_1 + R_2 \begin{bmatrix} \Box & \Box & \Box \\ \Box & \Box & \Box \end{bmatrix}$ $\begin{bmatrix} \Box & \Box & \Box \\ \Box & \Box & \Box \end{bmatrix} -\dfrac{1}{11}R_1 \begin{bmatrix} \Box & \Box & \Box \\ \Box & \Box & \Box \end{bmatrix}$ Write the equivalent system and solve. Answer:	

Example	Notes
2. Solve the system of equations using a matrix. $2x + 3y - z = 11$ $x + 2y + z = 12$ $3x - y + 2z = 5$ Answer:	

Helpful Hints

An augmented matrix is made up of two smaller matrices separated by a vertical line. The coefficients of the variable terms are placed to the left of the vertical line, and the constant terms are placed to the right.

In a matrix, all the numbers in any row or multiple of a row may be added to the corresponding numbers of another row. All the numbers in a row may be multiplied or divided by any nonzero number. Any two rows of a matrix may be interchanged.

Concept Check

1. Can you think of a situation where solving a system of linear equations using matrices would not be the best method? Explain.

Practice

Solve the systems of equations using matrices.

2. $-2x + 3y = 2$
   $-4x + y = -16$

3. $4x - 2y = 7$
   $3x + 2y = 7$

4. $x + 4y - 5z = 4$
   $3x - 8y + 7z = 9$
   $2x + 9z = 17$

More on Systems
Topic C.5 Cramer's Rule

Vocabulary
Cramer's rule   •   value of a 2×2 determinant   •   system of linear equations

1. _____ states that the solution to $\begin{array}{l} a_1 x + b_1 y = c_1 \\ a_2 x + b_2 y = c_2 \end{array}$ is $x = \dfrac{D_x}{D}$ and $y = \dfrac{D_y}{D}$,

   $D \neq 0$, where $D = \begin{vmatrix} a_1 & b_1 \\ a_2 & b_2 \end{vmatrix}$, $D_x = \begin{vmatrix} c_1 & b_1 \\ c_2 & b_2 \end{vmatrix}$, and $D_y = \begin{vmatrix} a_1 & c_1 \\ a_2 & c_2 \end{vmatrix}$.

Step-by-Step Video Notes
Watch the Step-by-Step Video lesson and complete the examples below.

Example	Notes
1. Solve the system by Cramer's rule. $-3x + y = 7$ $-4x - 3y = 5$ Find D. $D = \begin{vmatrix} \Box & \Box \\ \Box & \Box \end{vmatrix} = (\Box)(\Box) - (\Box)(\Box) = \Box$ Find D_x. Find D_y. Solve for x and y. $x = \dfrac{D_x}{D} = \dfrac{\Box}{\Box} = \Box$ and $y = \dfrac{D_y}{D} = \dfrac{\Box}{\Box} = \Box$ Answer:	

Example	Notes
3. Solve the system by Cramer's rule. $2x - y + z = 6$ $3x + 2y - z = 5$ $2x + 3y - 2z = 1$ Answer:	

Helpful Hints

The solution to $\begin{matrix} a_1x + b_1y + c_1z = k_1 \\ a_2x + b_2y + c_2z = k_2 \\ a_3x + b_3y + c_3z = k_3 \end{matrix}$ is $x = \dfrac{D_x}{D}$, $y = \dfrac{D_y}{D}$, and $z = \dfrac{D_z}{D}$, $D \neq 0$, where

$$D = \begin{vmatrix} a_1 & b_1 & c_1 \\ a_2 & b_2 & c_2 \\ a_3 & b_3 & c_3 \end{vmatrix}, \ D_x = \begin{vmatrix} k_1 & b_1 & c_1 \\ k_2 & b_2 & c_2 \\ k_3 & b_3 & c_3 \end{vmatrix}, \ D_y = \begin{vmatrix} a_1 & k_1 & c_1 \\ a_2 & k_2 & c_2 \\ a_3 & k_3 & c_3 \end{vmatrix}, \text{ and } D_z = \begin{vmatrix} a_1 & b_1 & k_1 \\ a_2 & b_2 & k_2 \\ a_3 & b_3 & k_3 \end{vmatrix}.$$

The value of the 2×2 determinant $\begin{vmatrix} a & b \\ c & d \end{vmatrix}$ is $ad - bc$.

Concept Check

1. Would you solve the system $\begin{matrix} x + y = 1 \\ -x + y = 3 \end{matrix}$ using Cramer's rule? Explain why or why not.

Practice
Solve the systems by Cramer's rule.

2. $\begin{matrix} 3x = 9 \\ -8x + 10y = -4 \end{matrix}$

3. $\begin{matrix} 4x + 6y = 48 \\ -x + 2y = 16 \end{matrix}$

4. $\begin{matrix} x + 2y + z = 5 \\ -x + 4y - 2z = 12 \\ -3x + y = 9 \end{matrix}$

Additional Topics
Topic D.1 Sets

Vocabulary

set • natural number • set intersection • set union • subset • empty set

1. A _____ is a collection of like objects called elements.

2. When all the elements of one set are contained in another set, the contained set is a _____ of the larger set.

Step-by-Step Video Notes
Watch the Step-by-Step Video lesson and complete the examples below.

Example	Notes
1 & 2. List all the elements in each set. Set X is the set of all natural numbers between 4 and 10 . Set Y is the set of all natural numbers between 2 and 7 , inclusive.	
3 & 4. Write in set-builder notation. $A = \{a,e,i,o,u\}$ Set B is the set of natural numbers between −6 and 0.	

Example	Notes
5. Find the intersection and union of the sets C and D. $C = \{$Odd numbers between 1 and 11, inclusive.$\}$ $D = \{$Multiples of 3 between 1 and 15, inclusive.$\}$	
6 & 7. Determine if the statement is true or false. Give the reason. $A = \{$odd numbers$\}, B = \{$integers$\}$, so $A \subseteq B$. $A = \{$multiples of 3$\}, B = \{$multiples of 4$\}$, so $A \subseteq B$.	

Helpful Hints

The range and domain of functions are often expressed using set-builder notation.

The intersection of two sets is the set of all the elements that are common to both sets. The union of two sets is the set of every element that is in either or both sets.

Concept Check

1. If $A \subseteq B$, what are $A \cap B$ and $A \cup B$?

Practice

Answer each based on the sets $A = \{2, 6, 10, 14...\}, B = \{4, 8, 12, 16...\}$, and $C = \{1, 3, 5, 7, ...\}$.

2. Find $A \cup B$. 4. Is $A \cup B \cup C = \{$natural numbers$\}$?

3. Find $A \cap B$. 5. Is $\{1, 2, 3, 4, 5\} \subseteq A \cup C$?

Additional Topics
Topic D.2 The Midpoint Formula

Vocabulary
midpoint   •   midpoint formula   •   length

1. The _____ of a line segment is the point on the line segment that is the same distance from each of the endpoints of the line segment.

Step-by-Step Video Notes
Watch the Step-by-Step Video lesson and complete the examples below.

Example	Notes
1. Find the midpoint of the line segment whose endpoints are $(1,4)$ and $(3,2)$. Let (x_1, y_1) represent one point and (x_2, y_2) represent the other. $(\Box, \Box)$ $x_1 \ y_1$ $(\Box, \Box)$ $x_2 \ y_2$ Substitute the values into the midpoint formula and simplify. $\dfrac{x_1 + x_2}{2} = \dfrac{\Box + \Box}{2} = \dfrac{\Box}{2} = \Box$ $\dfrac{y_1 + y_2}{2} = \dfrac{\Box + \Box}{2} = \dfrac{\Box}{2} = \Box$ Answer:	

Example	Notes
2–4. Find the midpoint of each line segment whose endpoints are the following points. $(-5,0)$ and $(9,-4)$ $\left(\dfrac{1}{2},-\dfrac{3}{4}\right)$ and $\left(-\dfrac{1}{2},\dfrac{1}{4}\right)$ $(20,-6.5)$ and $(-13,-7.3)$	

Helpful Hints

Do not confuse finding the midpoint of a line segment with finding the distance of a line segment.

The distance between two points is a length, whereas the midpoint of a line segment is a point on the line.

Concept Check

1. What is the relationship between the distance between an endpoint of a line segment and the midpoint of a line segment and the distance between the two endpoints of a line segment?

Practice

Find the midpoint of each line segment whose endpoints are the following points.

2. $(2,3)$ and $(4,5)$

4. $\left(\dfrac{2}{3},-\dfrac{2}{3}\right)$ and $\left(-\dfrac{2}{3},\dfrac{1}{3}\right)$

3. $(-7,0)$ and $(9,-6)$

5. $(-2,-3.4)$ and $(15,-7.2)$

Additional Topics
Topic D.3 Surface Area

Vocabulary
area   •   surface area   •   sphere

1. The _____ of a solid figure is the sum of the areas of all the surfaces.

Step-by-Step Video Notes
Watch the Step-by-Step Video lesson and complete the examples below.

Example	Notes
2. Find the number of surfaces and describe their shapes. Answer:	
4. Find the number of surfaces and describe their shapes. Answer:	
5. Find the surface area. 3 ft 8 ft 5 ft Answer:	

Example	Notes
8. The planet Mars has a diameter of approximately 6800 km. Find the surface area of Mars. (Assume the planet is a sphere.) Answer:	
9. A cardboard tube used for mailing posters is in the shape of a triangular prism, with each end shaped like an equilateral triangle. The sides of the triangle are each 5 inches, and the tube is 24 inches long. Find the surface area of the tube. Round your answer to the nearest hundredth. Answer:	

Helpful Hints
Surface area, like area, is measured in square units.

The surface area of a sphere with radius r can be found with the formula $A = 4\pi r^2$.

Concept Check
1. Determine the surface area, to the nearest hundredth, of the figure formed by a cylinder with height 13 meters and radius 4 meters that is topped by half of a sphere.

Practice
Find the number of surfaces and describe their shapes. Find the surface area to the nearest hundredth.

2.

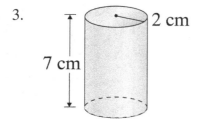

3 in

3.

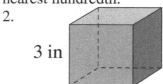

7 cm

2 cm

4. Saul wants to fold a sheet of cardboard into the shape of a triangular prism. The triangular ends of the prism are right triangles, 4 in. high by 3 in. wide. The prism will have a length of 10 in. How much cardboard is needed to make the prism?

Additional Topics
Topic D.4 Synthetic Division

Vocabulary
synthetic division   •   quotient   •   divisor   •   dividend

1. Dividing a polynomial by a binomial can be made more efficient with a process called _____.

Step-by-Step Video Notes
Watch the Step-by-Step Video lesson and complete the examples below.

Example	Notes
1. Divide using synthetic division. $$\left(3x^3 + 7x^2 - 4x + 3\right) \div \left(x + 3\right)$$ Turn the first three numbers in the bottom row into the quotient. The last number in the bottom row is the remainder. Put this over the divisor to complete the answer. Answer:	

Example	Notes
2. Divide using synthetic division. $$\left(3x^4 - 21x^3 + 31x^2 - 25\right) \div \left(x - 5\right)$$	

Helpful Hints

"Missing terms" refers to monomial terms with degrees between the degrees of the terms in the polynomial being divided. For example, if the polynomial being divided is $3x^3 + 1$, the missing terms are $0x^2$ and $0x$.

Be sure to include zeros for the missing terms when using synthetic division.

Concept Check
1. What is the disadvantage of synthetic division?

Practice
Divide using synthetic division.

2. $\left(x^3 + 4x + 4\right) \div \left(x + 1\right)$

4. $\left(x^4 + 3x^3 + 5x^2 + 18x + 9\right) \div \left(x + 3\right)$

3. $\left(x^5 + 4x^2 + x\right) \div \left(x - 2\right)$

5. $\left(x^3 - 8\right) \div \left(x - 2\right)$

Additional Topics
Topic D.5 Balancing a Checking Account

Vocabulary
balancing a checkbook   •   credits   •   debits

1. _____ are any amounts that are deducted from your checking account.

2. _____ are amounts that are added to your checking account.

Step-by-Step Video Notes
Watch the Step-by-Step Video lesson and complete the examples below.

Example

1. As of 7/11, Alex Murgaton had a balance of $1657.38 in his checking account. On 7/12, Alex wrote check #177 to Shop Market for $122.96. On 7/13, he used his debit card at Chelsea's Cinema for $16 and at Bob's Convenience Store for $10.97. On 7/14, his weekly check of $590.27 from work was directly deposited into his account, and he withdrew $60 from an ATM. On 7/15, his automatic payments of $30.02 to the Electric Company, $17.76 to the Gas Company, and $650.00 to Sky High Apartments were paid. Record the credits and debits in Alex's check register and then find his ending balance.

CHECK REGISTER *Alex Murgaton*							20 *11*	
CHECK NO.	DATE	DESCRIPTION OF TRANSACTION	PAYMENT/ DEBIT (−)	✓	DEPOSIT/ CREDIT (+)		BALANCE $ *1657*	*38*
177	*7/12*	*Shop Market*	*122* *96*	✓			*1534*	*42*

Notes

Example

2. Morgan O'Malley calculated her checkbook balance, but it does not match her bank's records. Her bank states that as of 4/8, she should have $600.53 in her checking account. Look at Morgan's check register and identify and fix any errors that she might have made. If she made no errors, then indicate that the bank must have made an error.

CHECK NO.	DATE	DESCRIPTION OF TRANSACTION	PAYMENT/ DEBIT (−)	✓	DEPOSIT/ CREDIT (+)	BALANCE $568 17
303	4/2	Pet Market	44 02	✓		524 15
	4/3	Direct Deposit		✓	988 15	1512 30
	4/5	River Place Apartments		✓	767 99	2280 80
304	4/5	Violin Lessons	30 00	✓		2250 29
	4/8	Al's Food Market (debit card)	113 78	✓		2126 51

CHECK REGISTER Morgan O'Malley — 20 12

Notes

3. Balance the monthly statement for Lorence Seafood using the restaurant's check register and bank statement.

CHECK NO.	DATE	DESCRIPTION OF TRANSACTION	PAYMENT/ DEBIT (−)	✓	DEPOSIT/ CREDIT (+)	BALANCE $2607 33
268	2/2	Shoreline Fishery	613 07	✓		
269	2/10	Tia Sullivan	430 00	✓		
270	2/10	Moe Cormorant	430 00	✓		
271	2/10	Evan Enriquéz	620 00			
	2/14	Deposit		✓	1770 00	
272	2/20	Raymund Property Management	575 00	✓		
273	2/24	Wegstaff Appliance Repair	447 74	✓		
	3/1	Deposit		✓	1300 00	

CHECK REGISTER Lorence Seafood — 20 12

(Continued on the next page.)

Bank Statement: LORENCE SEAFOOD	2/1/2012 to 2/29/2012

Beginning Balance $2607.33
Ending Balance $1875.02

Checks cleared by the bank

#268	$613.07	#270*	$430.00	#273	$447.74
#269	$430.00	#272	$575.00		

*Indicates that the next check in the sequence is outstanding (hasn't cleared).

Deposits

2/14 $1770.00

Other withdrawals

Service charge $6.50

CHECKING RECONCILEMENT   This form is provided to assist you in balancing your checking account.

List checks outstanding* not charged to your checking account

Period ending **2/29 , 20 12**

CHECK NO.	AMOUNT
271	620 00
TOTAL	620 00

*and ATM withdrawals

1. Check Register Balance — $ **2561.52**

   Subtract any charges listed on the bank statement which you have not previously deducted from your balance.   − $ **6.50**

   Adjusted Check Register Balance $

2. Enter the ending balance shown on the bank statement. $ **1875.02**

3. Enter deposits made later than the ending date on the bank statement.   + $ **1300.00**
    + $
    + $

TOTAL (Step 2 plus Step 3) $

4. In your check register, **check off** all the checks paid. In the area provided to the left, **list** numbers and amounts of all outstanding checks and ATM withdrawals.

5. Subtract the total amount in Step 4.   − $

6. This adjusted bank balance should equal the adjusted Check Register Balance from Step 1. $

Notes

Helpful Hints

If you prefer to keep track of your records on your computer, create a spreadsheet that includes columns that match the checkbook register headings.

Concept Check

1. Why is it important to balance your checkbook each month?

Practice

2. As of 3/12, Dale Archer had a balance of $2485.32 in his checking account. On 3/15, Dale used his debit card at the Gas'n'Bubbles gas station and car wash for $36.61. On 3/21, Dale wrote check #353 to Phil's Pharmacy for $5.99. That night, he went on a date to Moovie's Cinema and Steakhouse, where he used his debit card again and paid $51.32. On 3/29, his automatic payments of $81.84 to the Cable Company, and $42.25 to the Electric Company were paid. On 3/31, Dale wrote check #355 for $750.00 to the Laketown Realty Company. On 4/1, Dale's paycheck of $625.25 was directly deposited into his account. Record the credits and debits in Dale's check register and find his ending balance.

CHECK REGISTER						20 ____	
CHECK NO.	DATE	DESCRIPTION OF TRANSACTION	PAYMENT/ DEBIT (−)	✓	DEPOSIT/ CREDIT (+)	BALANCE $	

3. Joe Grey found that the bank's record of his checkbook balance does not match the balance he calculated. His bank states that as of 11/12, he should have $7245.32 in his checking account. Look at Joe's check register and identify and fix any errors he may have made. If Joe made no errors, then indicate that the bank must have made an error.

CHECK REGISTER	Joe Grey					20 14	
CHECK NO.	DATE	DESCRIPTION OF TRANSACTION	PAYMENT/ DEBIT (−)	✓	DEPOSIT/ CREDIT (+)	BALANCE $ 6818 66	
	11/6	Birthday – Grandma		✓	60 00	6878	66
	11/7	Cable Company	42 61	✓		6836	05
	11/8	Bill's Supermarket (debit card)	32 81	✓		6803	24
	11/11	Pay Check		✓	450 00	7253	24
111	11/12	Electric Company	15 92	✓		7237	32

4. Balance the monthly statement for Tim's Hobby Hangar using the store's check register and bank statement.

CHECKING RECONCILEMENT   This form is provided to assist you in balancing your checking account.

Period ending: 4/30, 20 14

List checks outstanding* not charged to your checking account		
CHECK NO.	AMOUNT	
195	130	21
202	240	16
TOTAL		

*and ATM withdrawals

1. Check Register Balance	$ 6843.25
Subtract any charges listed on the bank statement which you have not previously deducted from your balance. —	$ 34.71
Adjusted Check Register Balance	$
2. Enter the ending balance shown on the bank statement.	$
3. Enter deposits made later than the ending date on the bank statement. +	$ 860.00
+	$ 112.91
+	$
TOTAL (Step 2 plus Step 3)	$
4. In your check register, **check off** all the checks paid. In the area provided to the left, **list** numbers and amounts of all outstanding checks and ATM withdrawals.	
5. Subtract the total amount in Step 4. —	$
6. This adjusted bank balance should equal the adjusted Check Register Balance from Step 1.	$

Name: _____   Date: _____

Instructor: _____   Section: _____

Additional Topics
Topic D.6 Determining the Best Deal When Purchasing a Vehicle

Vocabulary
total cost   •   purchase price   •   down payment   •   amount financed

1. A(n) _____ is the initial amount paid at the time of purchase.

2. The _____ is the total amount of money spent.

Step-by-Step Video Notes
Watch the Step-by-Step Video lesson and complete the examples below.

Example	Notes

1. Scott purchased a truck that was on sale for $29,999 in a city that has a 6.5% sales tax and a 2% license fee.
  a. Find the amount of sales tax and license fee.
  b. Scott also bought an extended warranty for $1650. Find the purchase price of the truck.
 	
  Answer:	

2. The purchase price of a car Lisa wants to buy is $15,999. In order to qualify for the loan on the car, Lisa must make a down payment of 20% of the purchase price.
  a. Find the amount of the down payment.
  b. Find the amount financed.
 	
  Answer:	

Example	Notes
3. James went to two dealerships to find the best deal on a sports car he plans to purchase. From which dealer should James buy the car so that the total cost of the car is the least expensive?	

Dealership 1	Dealership 2
Purchase price: $36,999	Purchase price: $36,999
Financing option: 3% financing with $5000 down payment	Financing option: 5% financing with no down payment
Monthly payments: $686.65 per month for 48 months	Monthly payments: $647.48 per month for 60 months

Answer:

Helpful Hints

There is more to buying a car than just the price. Taxes, license fees, extended warrantees, and interest on a loan all add to the total cost of the vehicle.

Concept Check

1. How can increasing the amount of a down payment reduce the total cost of a car?

Practice

2. Sarah purchased a car that was on sale for $12,450 in a state that has a 7% sales tax and a 3% license fee. Find the amount of sales tax and license fee Sarah paid.

3. The purchase price of a truck Paul wants to buy is $14,399. In order to qualify for a loan, Paul must make a down payment of 15% of the purchase price.
   a. Find the amount of the down payment.
   b. Find the amount financed.

4. When shopping for a car, Evan went to two dealerships to find the best deal. Dealership 1 could finance the car at monthly payments of $345.72 for 60 months. Dealership 2 could finance the car at monthly payments of $320.42 for 48 months, but with a $4000 down payment. Which of the dealerships has the lower total price?